IMAGES
of America

THE TENNESSEE COPPER COMPANY
1899–1970

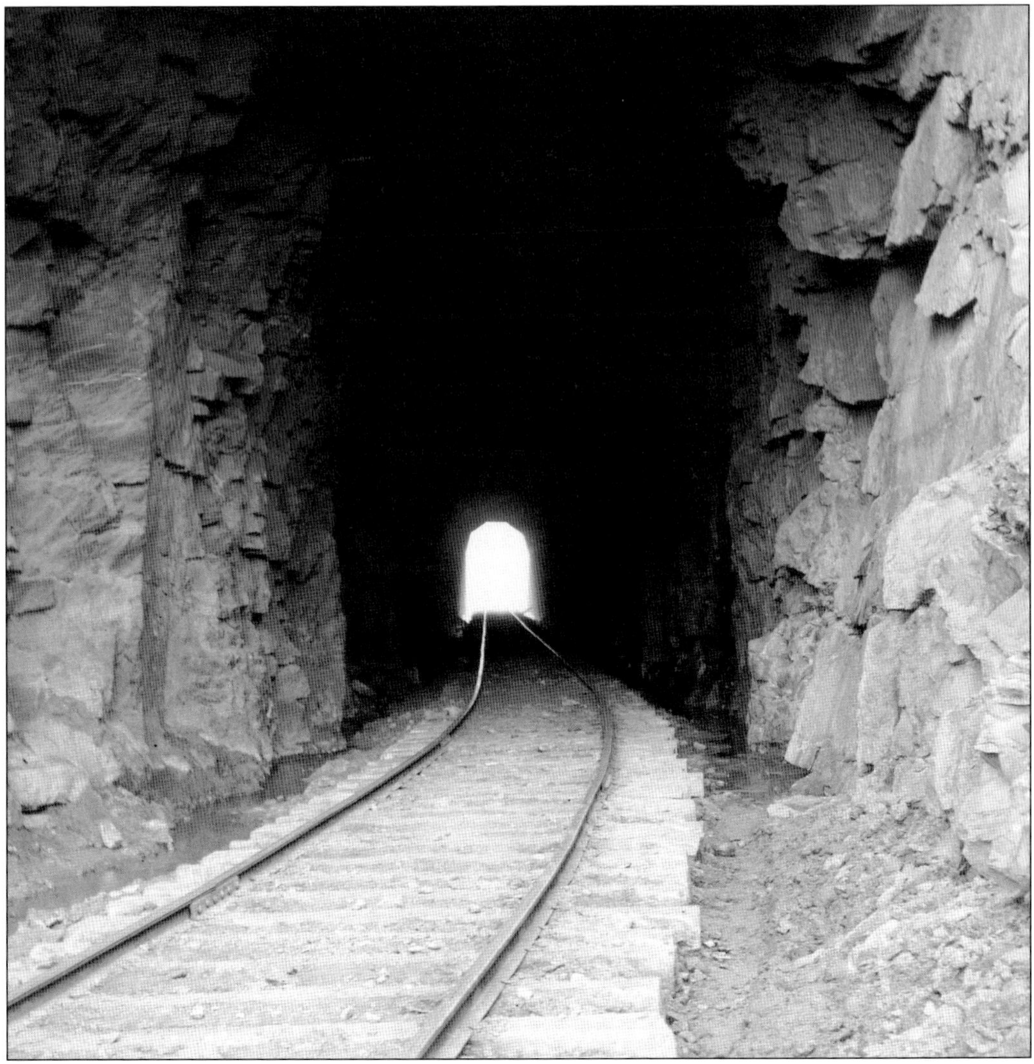

Railway Tunnel, 1903. In 2010, more than 200 historical photographs from the albums of Tennessee Corporation president Emory Hammond Westlake were donated to the Ducktown Basin Museum by his grandson Wes Ackley. They were meticulously captioned, with dates ranging from the earliest days of the Tennessee Copper Company to its expansion during and shortly after World War I. The oldest photographs are thought to have been taken by Westlake himself as a young engineer starting out in the mining industry. Many of those images are being published here for the first time. One of the earliest is this one, featuring a tunnel along the route of the Tennessee Copper Company railway in 1903. (Courtesy of the Ducktown Basin Museum.)

On the Cover: In 1960, drillers at Cherokee Mine prepare to lower a drilling jumbo into the shaft for continuing development of the last Tennessee Copper Company underground mine. Cherokee Mine began ore production in 1961, continued in operation through successive managements, and was the last of the Ducktown district mines to close, in 1987. (Courtesy of the Ducktown Basin Museum.)

IMAGES
of America

The Tennessee Copper Company 1899–1970

Harriet Frye

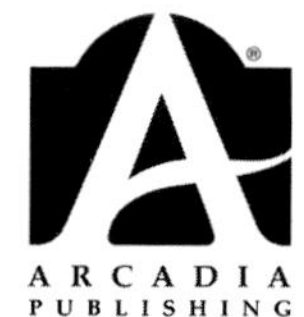

ISBN 978-1-4671-0764-8

Published by Arcadia Publishing
Charleston, South Carolina

Printed in the United States of America

Library of Congress Control Number: 2021946112

For all general information, please contact Arcadia Publishing:
Telephone 843-853-2070
Fax 843-853-0044
E-mail sales@arcadiapublishing.com
For customer service and orders:
Toll-Free 1-888-313-2665

Visit us on the Internet at www.arcadiapublishing.com

Dedicated to the men and women of the Tennessee Copper Company, their families, and their descendants, past and present.

Contents

Acknowledgments 6

Introduction 7

1. 1899–1949 9
2. The 1950s 43
3. The 1960s 77
4. Life after Work 101

Bibliography 126

The Ducktown Basin Museum 127

Acknowledgments

In 2019, the Ducktown Basin Museum discovered a figurative gold mine of its own. Stuffed into the drawers of a donated file cabinet were several thousand professional photographs documenting the pictorial history of Tennessee Copper Company and its successor companies. Many of them were indexed to related stories in the company newsletter, *TC Topics*. The contents of that cabinet were the inspiration for this book and the primary source of the images in the last three chapters.

First, then, I need to thank Frank Russell of Glenn Springs Holdings for locating that file cabinet and Buddy Haynes of Copperhill Industries for donating it; current director Sarah Mickens and the museum staff for allowing me unrestricted access to the files; and former museum director Ken Rush, who not only put those photographs in my hands in the first place, but also reviewed the final text of this book for accuracy. After 30 years as director prior to his retirement in 2020, Ken knows far more about the workings of the Tennessee Copper Company than I ever will.

I also need to thank the late Wes Ackley, grandson of Tennessee Corporation president E.H. Westlake, for donating his grandfather's photograph albums to the museum. Without those remarkable historic images (credited as "EHW"), the first chapter of this book would have been only about half as long.

Angel Prohaska, my title manager at Arcadia, not only guided me through the production process but also cheerfully answered what she must have thought were some very peculiar questions. Thanks, Angel. I hope the result was worth it.

I could never have written this book without the meticulous research of the late R.E. Barclay, whose books on local history are still the definitive source of information on early mining in the Ducktown district; and I would never even have attempted it without the contributions of now-deceased *TC Topics* editors Bill Hyde and Don Sisson, who took and annotated most of the photographs in the last three chapters.

Unless otherwise noted, all the images in this book are from the historical photograph archives of the Ducktown Basin Museum.

INTRODUCTION

In 1897, as an agent for the New York City firm of Lewisohn Brothers, consulting engineer J. Parke Channing secured options on all the available mineral properties in the southeastern Tennessee district known to geologists as the Ducktown Basin. When exploratory drilling proved the feasibility of reopening the ore bodies of the district's pre–Civil War Burra Burra and London Mines, which had lain idle since the 1870s, Lewisohn Brothers moved forward with the purchase of the optioned properties. They also acquired the facilities and leases of the Pittsburgh & Tennessee Copper Company, which had briefly reopened the old Polk County Mine a few years earlier. With these acquisitions, Lewisohn Brothers came into possession of virtually all the Ducktown district mining properties other than those already owned by the smaller Ducktown Sulphur, Copper & Iron Company Ltd. (DSC&I), which had begun operations in 1891.

The Tennessee Copper Company (TCC) was incorporated on April 25, 1899. The sinking of new shafts at Burra Burra and London Mines and the dewatering and expansion of the existing Polk County shaft were begun almost at once. By mid-May 1900, the company had broken ground for a smelter and other plant facilities on a hill overlooking the small town of McCays, Tennessee, which would be renamed Copperhill in 1907.

In 1901, open heap roasting of copper ore, the first step in the smelting process, began at Burra Burra and Polk County Mines. This was also where the trouble began—or, to be more accurate, where it began for the second time. Open roasting, also employed by earlier mining companies, created a sulfurous smoke that spread across the surrounding areas like a dense fog, hugging the ground and destroying vegetation in its path. The roast heaps also devoured huge quantities of timber. With the tall trees gone, the sulfur fumes functioned as acid rain, eventually killing off the undergrowth and exposing the denuded ground to constant erosion. By the time TCC and DSC&I began operations, much of the land surrounding the original roasting and smelting facilities was already a red-clay desert, its vegetation long gone and its topsoil washed away into the Ocoee River.

Not surprisingly, the renewal of roasting and smelting operations set off a furor in the nearby Tennessee and Georgia communities that led to a series of lawsuits and injunctions against both companies for sulfur smoke damage. The legal challenges were ultimately dropped, but only after several years had passed and only after both companies had made extensive changes to their operations.

By 1903, both TCC and DSC&I had begun to convert to pyritic smelting, which ended the need for open roasting. By the end of 1904, the last of the roast yards had been closed. Unfortunately, the smelter was still emitting sulfurous gases, the tall chimneys were dispersing the fumes over a wider area, and many residents claimed that the transition had actually made the problem worse. At this point, TCC made a fateful decision. Channing, now president of the company, convinced the directors to gamble on the construction of a large lead chamber acid plant adjoining the smelter at McCays. The use of copper blast-furnace gases to produce sulfuric acid had never been tried before on a commercial scale; the chamber plant would be a last-ditch effort to keep the mines and smelters in operation, and virtually no one expected it to be a serious source of profit. In fact, it was to change the course of TCC history.

Before the plant was even completed, orders for sulfuric acid began to arrive. By the time the first tank cars of acid were shipped out in January 1908, a surprised TCC management had already contracted to sell its entire output for the year. In 1909, the company began construction on a second lead chamber unit that would double the size of the existing plant, making it the largest sulfuric acid plant in the world.

In the 1920s, both TCC and DSC&I began to move ahead with plans to convert from smelting to milling, a process that allowed for the efficient recovery of other components from the raw ore. In many ways, the two companies were following parallel tracks, although TCC would remain the larger and more profitable of the two. They continued to operate virtually side by side until 1936, when TCC purchased the properties of DSC&I's struggling successor company, the Ducktown Chemical & Iron Company (DC&I), and integrated them into the TCC operations. The consolidation made TCC the largest US copper mining company east of the Rocky Mountains and south of Lake Superior.

Meanwhile, at Copperhill, the by-products of acid production and milling had become the basis for on-site manufacturing of a growing list of industrial products. By mid-century, local high-school students were participating in essay contests on the subject "What the Chemical Industry Means to My Family and Me," and production of industrial chemicals and metals had become the company's primary function.

In November 1962, to the dismay of many employees and other local residents, an agreement was reached by TCC's parent company, Tennessee Corporation, for a pooling of interests with Cities Service Company. The merger became official on June 14, 1963. On January 1, 1970, Tennessee Corporation was absorbed into Cities Service, and the Tennessee Copper Company officially ceased to exist.

The former TCC operations changed hands several more times over the next four decades. Facing increased production costs, decreased copper prices, and a rapidly changing economy, the successor companies struggled to readjust. Over time, it became increasingly clear that what had worked for more than three-quarters of a century was no longer working—and that there was no longer any feasible way to make it work. Between 1971 and 1987, the remaining underground mines were closed and flooded. By 2000, what remained of TCC's once-massive operation required a workforce of just 40 employees. A decade later, even those jobs had vanished, and the largest (and ultimately most successful) environmental cleanup and reclamation project in Tennessee history was underway.

Today, the primary industry of the Ducktown district, now better known as the Copper Basin, is tourism. The river and the creeks run clear again, outdoor activities draw thousands of visitors every year, and the bare red hills of yesterday are nowhere to be seen. What often surprises newcomers and visitors is that some older residents still miss those red hills. They had been there for so long that they became part of the collective consciousness, and the flourishing vegetation that covers them now seems vaguely out of place.

The largely intact Burra Burra Mine complex is now the home of the Ducktown Basin Museum. The General Office building still overlooks the city of Copperhill, and a scant few of the abandoned plant facilities still stand nearby. The lone headframe of Central Shaft still rises above the highway between Copperhill and Ducktown.

Otherwise, of the once-mighty Tennessee Copper Company, virtually all that remains are the memories.

One

1899–1949

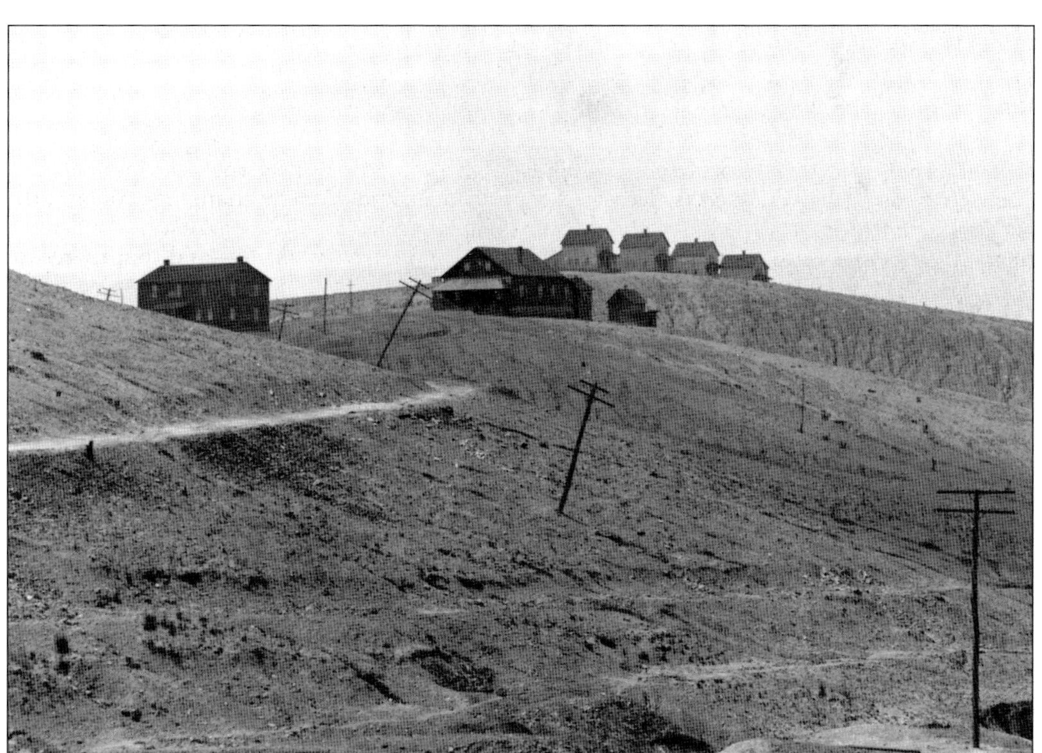

TEMPORARY HEADQUARTERS, 1903. While TCC completed work on its new headquarters complex, the managerial staff moved into the former offices of the Pittsburgh & Tennessee Copper Company at Polk County Mine. The site also included a supply house, a commissary, and management housing. It was in use until 1904, when the new General Office and management housing were completed at McCays. (EHW.)

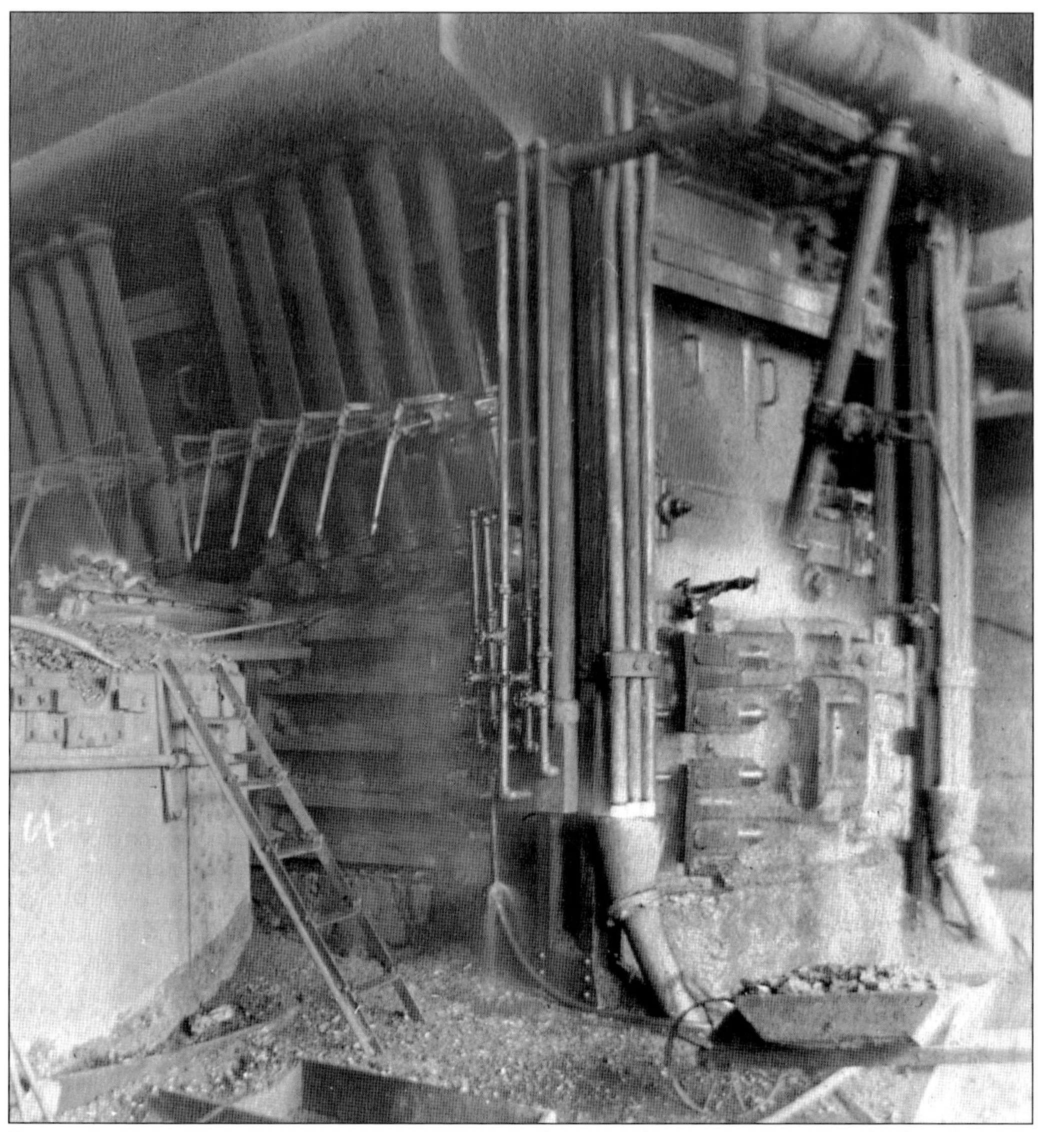

Blast Furnace, Early 1900s. In 1900, an Iowa-born smelting engineer named Edward Oscar Boak was hired to help set up TCC's first smelting operation. He remained on the job for five years and is thought to have taken this photograph of an early blast furnace around 1904 or 1905. The furnace in this photograph, one of the first three to be installed at the smelter, was 15 feet wide. The furnaces were fed from the top through openings in the charge floor on the upper level. By 1906, after the expansion of the original smelter, TCC had seven blast furnaces in operation. With advances in technology over the next few decades, including the company's conversion to milling for the initial processing of ore, the blast furnaces were gradually replaced by reverberatory furnaces. The last blast furnace was permanently retired in 1947.

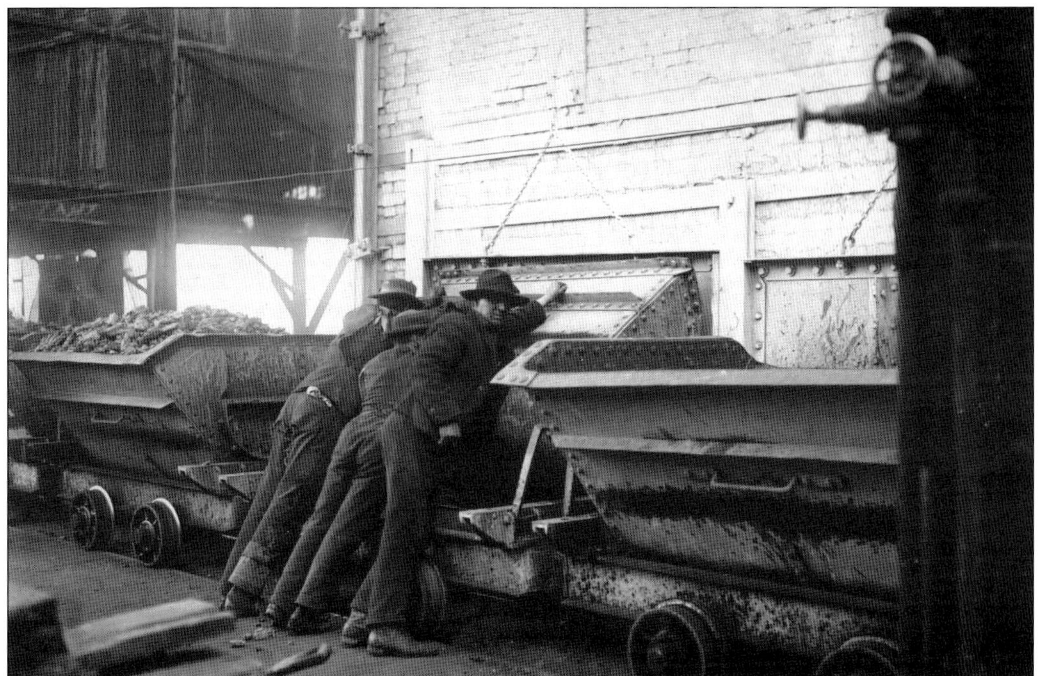

Charge Floor, 1904. In the pyritic smelting process, the raw ore, mixed with coke and other materials to aid combustion, was fed manually into the blast furnaces. This process, known as charging, was the first step in the smelting operation. The charge floor was on the upper level of the smelter just above the furnaces, and small dump cars ran along a track from the exterior ore bins to the charging level. Above, three employees feed the load from one of the dump cars into the furnace. Below, a wider view shows the extent of the charging floor in 1904, when three furnaces were in operation. (Both, EHW.)

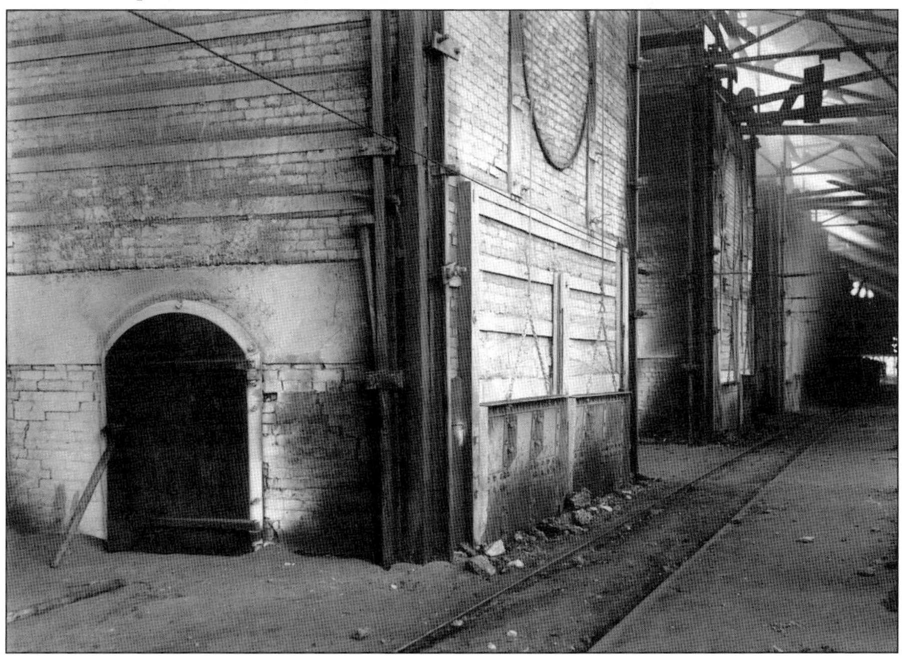

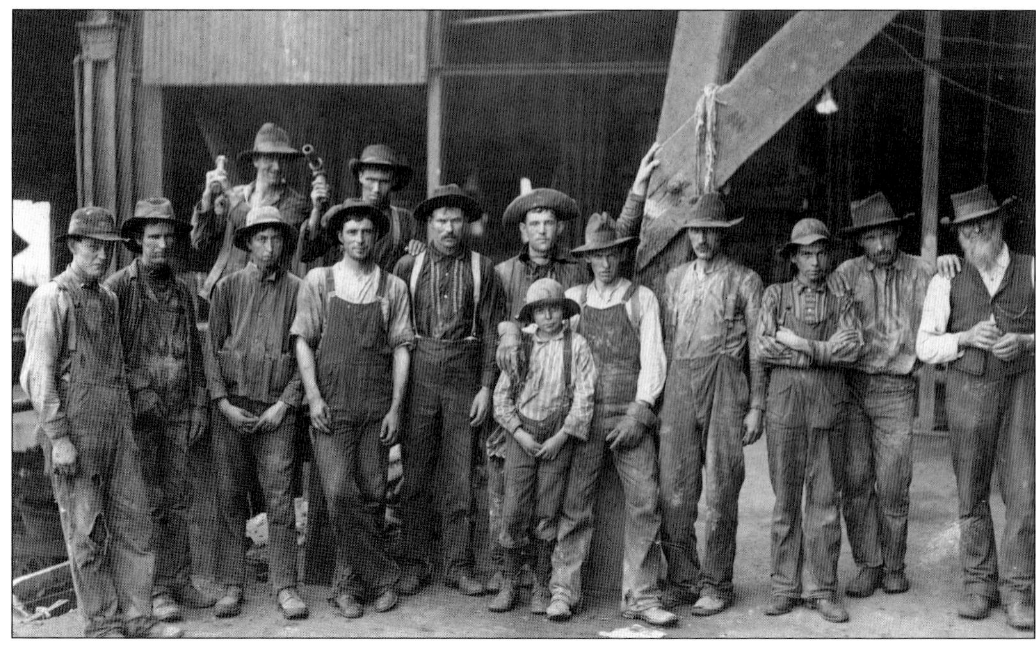

Settler Floor and Converters, Early 1900s. On the settler floor (above, in 1904), the melted ore from the furnaces was drawn into bins, where the copper matte settled to the bottom and could be drawn out for re-smelting. The separated matte was then broken up and fed back into the furnaces. The result of the second smelting was a mixture that contained about 30 percent copper. The final mixture was transferred by ladles to converters (below), which converted it into an almost pure form of Bessemer pig copper for shipment to refineries in the United States and Europe. (Above, EHW.)

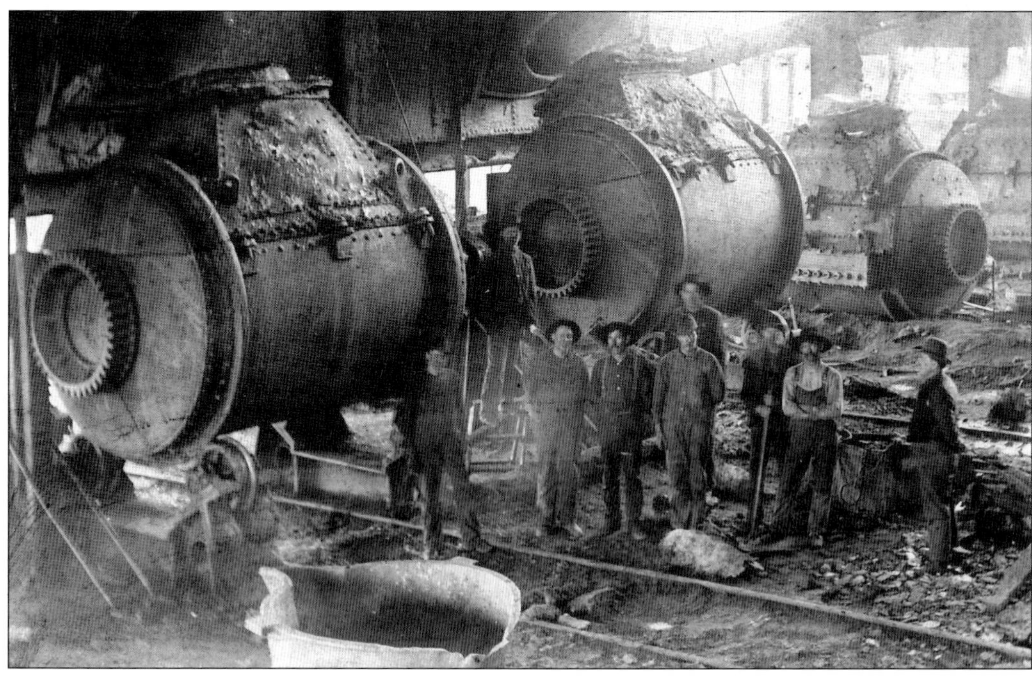

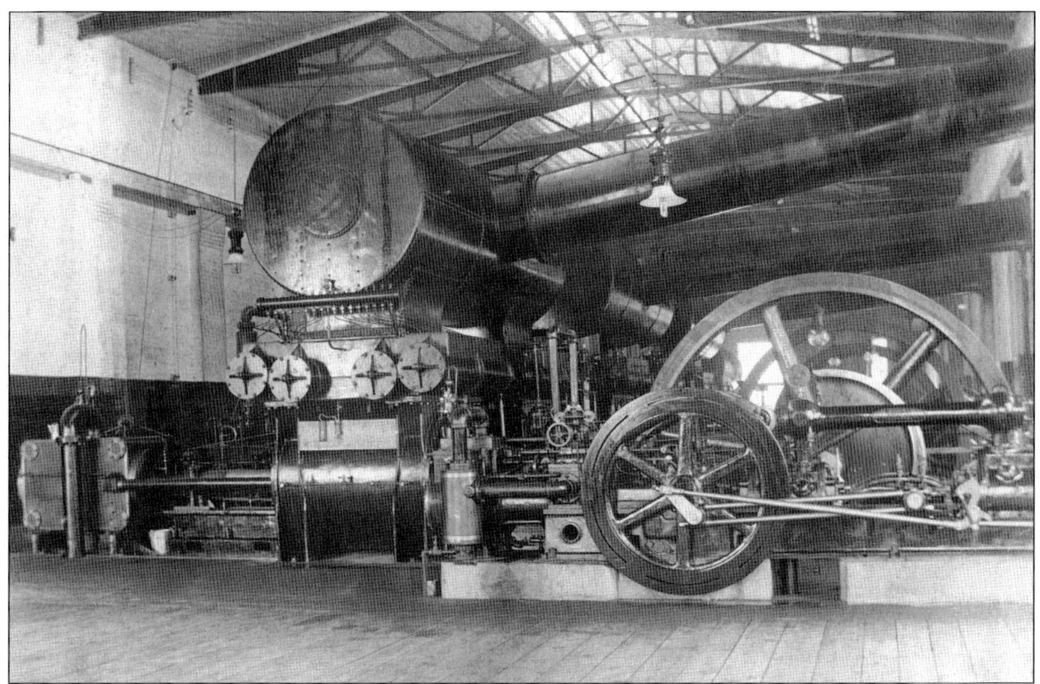

Powerhouse at Smelter. The powerhouse/boiler room at the TCC smelter originally included two Nordberg condensing blowing engines for blast furnaces, one blower for converters, four coal-fired boilers, and an electric shop to power the smelter. It also powered locomotives for charging cars and slag cars, as well as all building lights. A third Nordberg engine (above), two Murphy boilers, mechanical stokers, and a complete electric lighting plant were added in 1904. The interior was photographed around 1904 by Edward Oscar Boak (above) and in 1906 by E.H. Westlake (below).

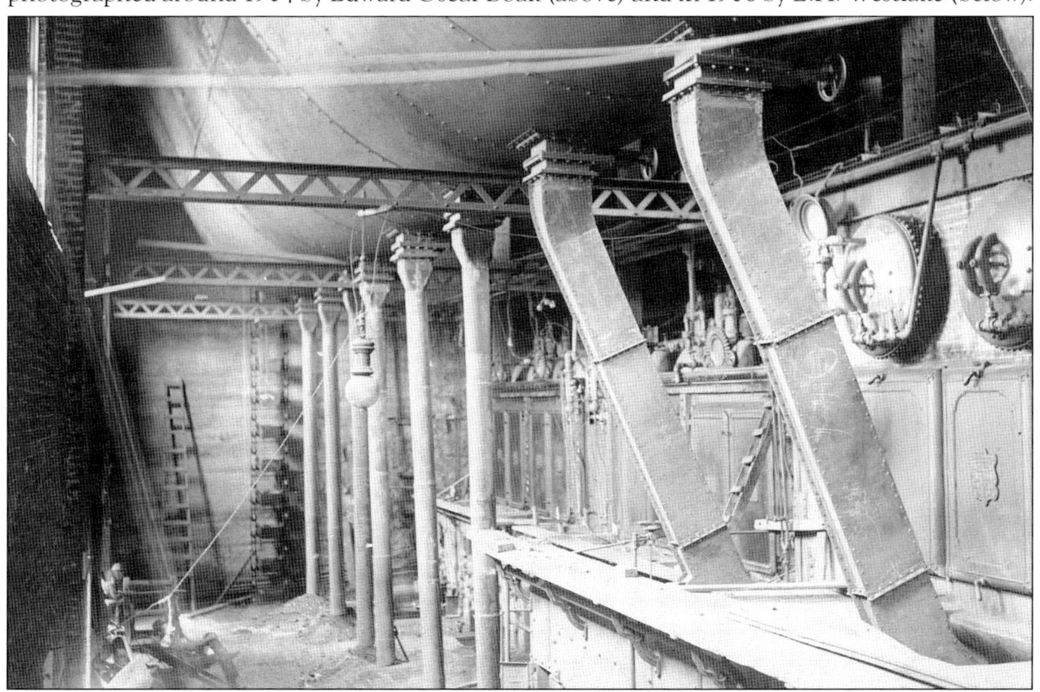

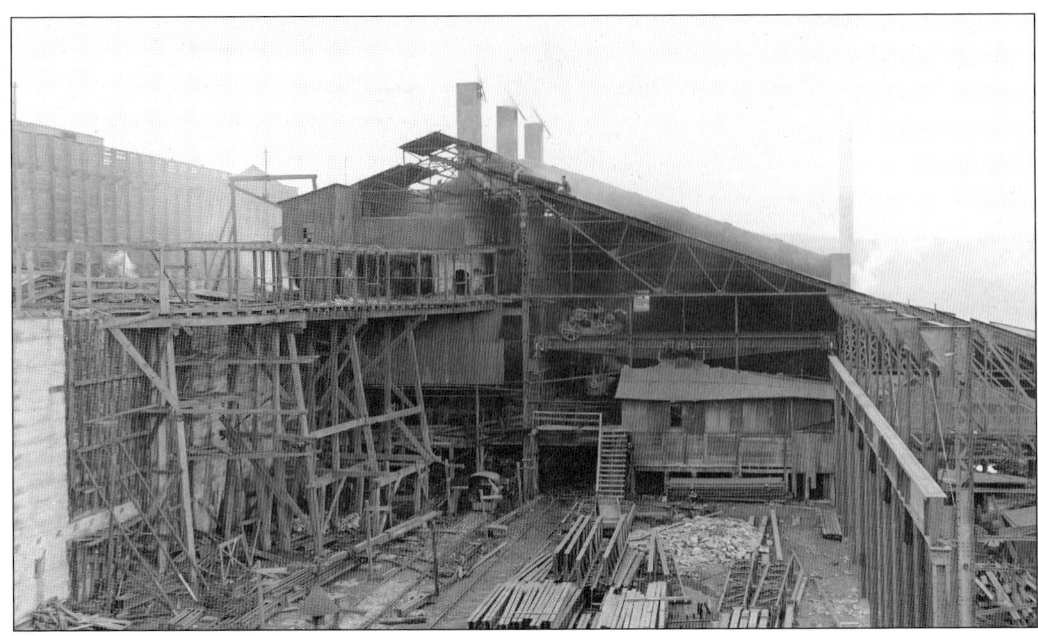

Smelter Extension, 1905. By the end of 1904, TCC had discontinued the open roasting process and converted entirely to pyritic smelting. The changeover necessitated the enlargement of the smelter at McCays to include four new blast furnaces, which were located west of the three existing furnaces whose chimneys can be seen above. At 22 feet wide, the new additions were the largest copper blast furnaces in the United States. Among the heavy equipment needed for the construction process, which continued into 1906, was the Morgan steam-powered overhead crane that can be seen in more detail below. (Both, EHW.)

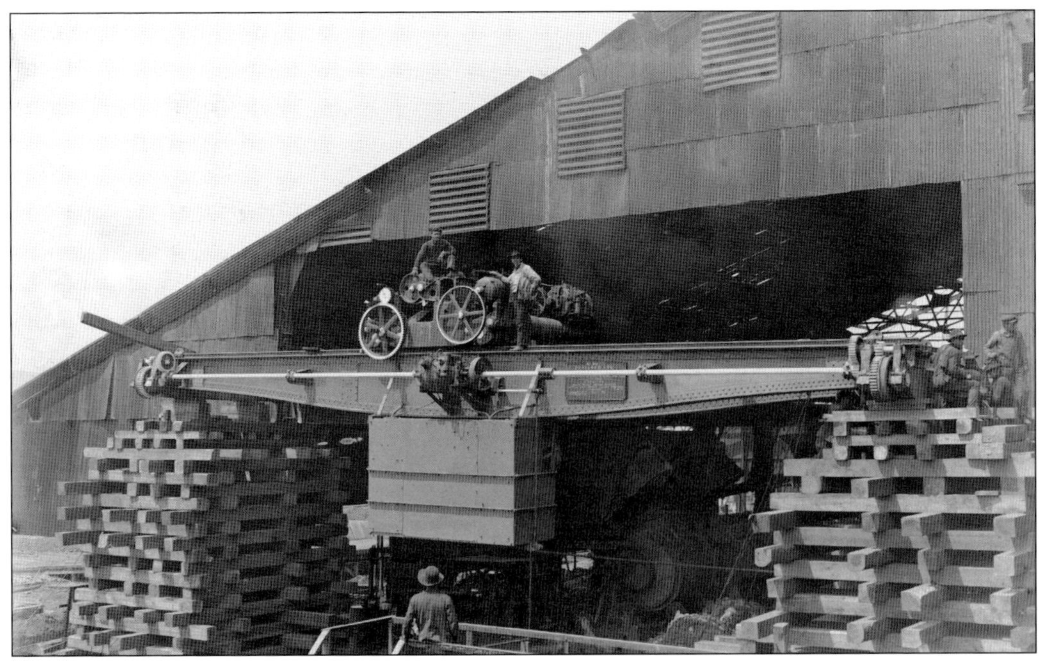

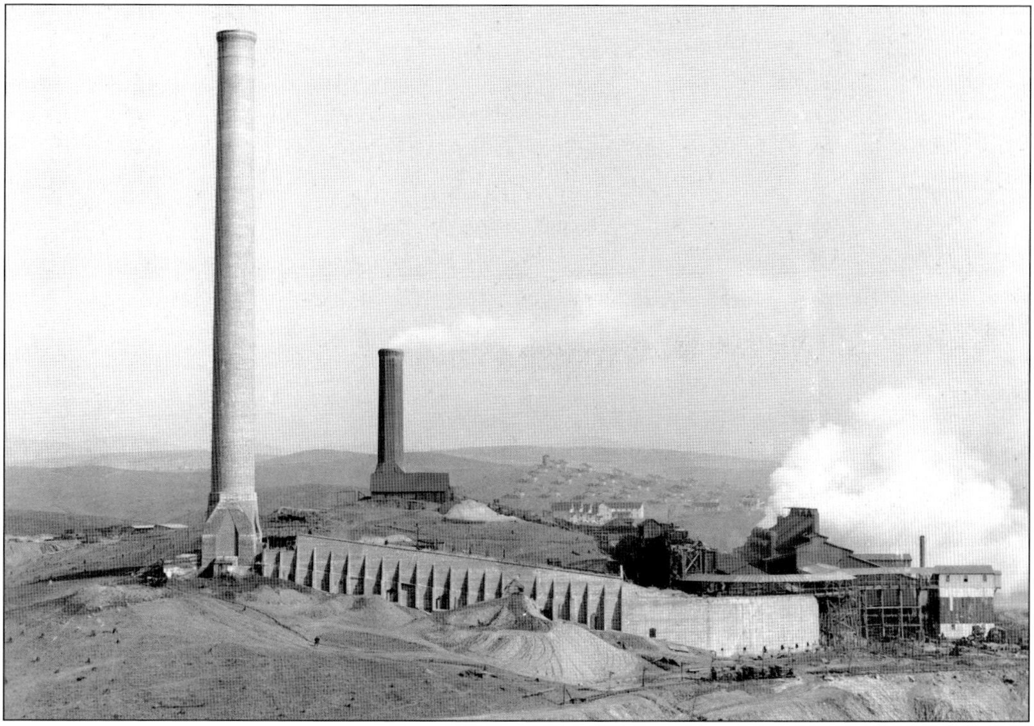

NEW SMELTER CHIMNEY, 1906. As part of the smelter extension project, a new 325-foot chimney (left) was erected to replace the original 150-foot chimney (center), which is still in operation in this photograph. Later that year, the older chimney was demolished, and the hilltop was leveled to prepare the site for TCC's first sulfuric acid plant. (EHW.)

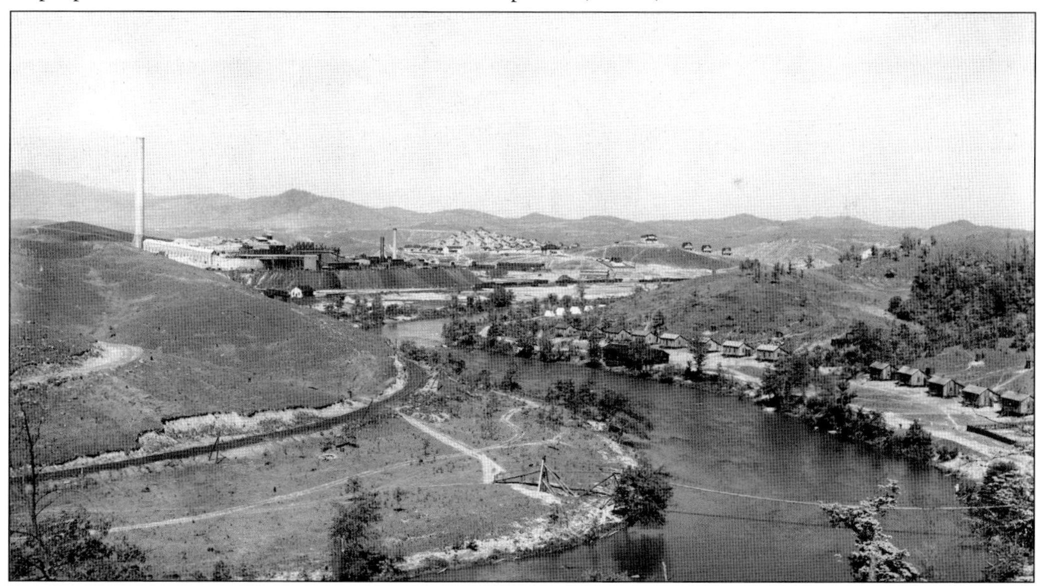

TCC PLANT AND ENVIRONS, 1906. In this slightly later photograph, smoke can be seen coming from the new blast furnaces (left), and the acid plant site has been leveled. In addition to the company housing near the plant, small houses line one bank of Fightingtown Creek, which branched off the Ocoee River just across from the plant site. (EHW.)

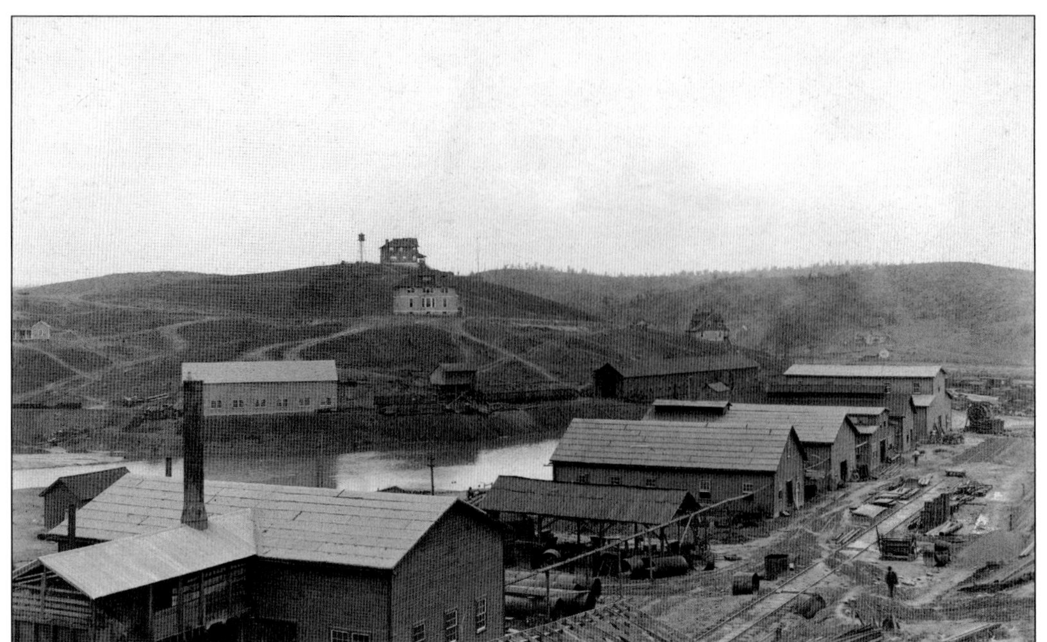

Shops, 1905. In early 1905, TCC completed construction of a machine shop, a carpenter/pattern shop, a plate shop, a blacksmith shop, a foundry, two railcar repair shops, and a matte sampling building. In this photograph, the general manager's house (at the top of the hill) and the General Office (just below it) overlook the row of newly completed shops. The smelter superintendent's house is farther downhill on the right. (EHW.)

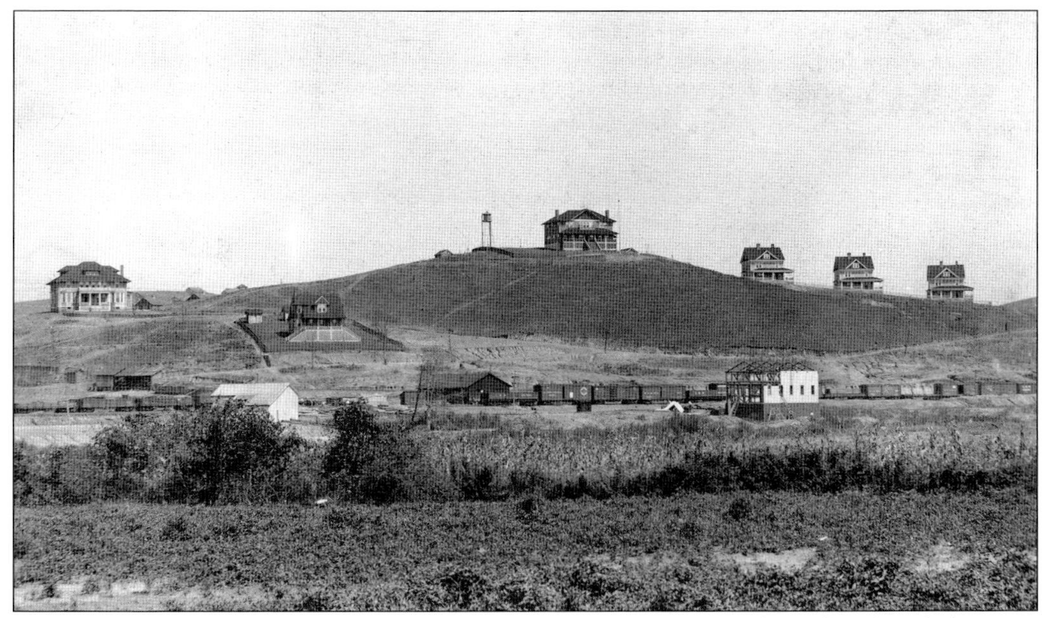

Managerial Housing, 1905. In this view from a vantage point near the railroad yard, the TCC General Office is at left, with the smelter superintendent's house to the right and slightly below. The general manager's house is at the top of the hill near a row of three identical executives' houses. (EHW.)

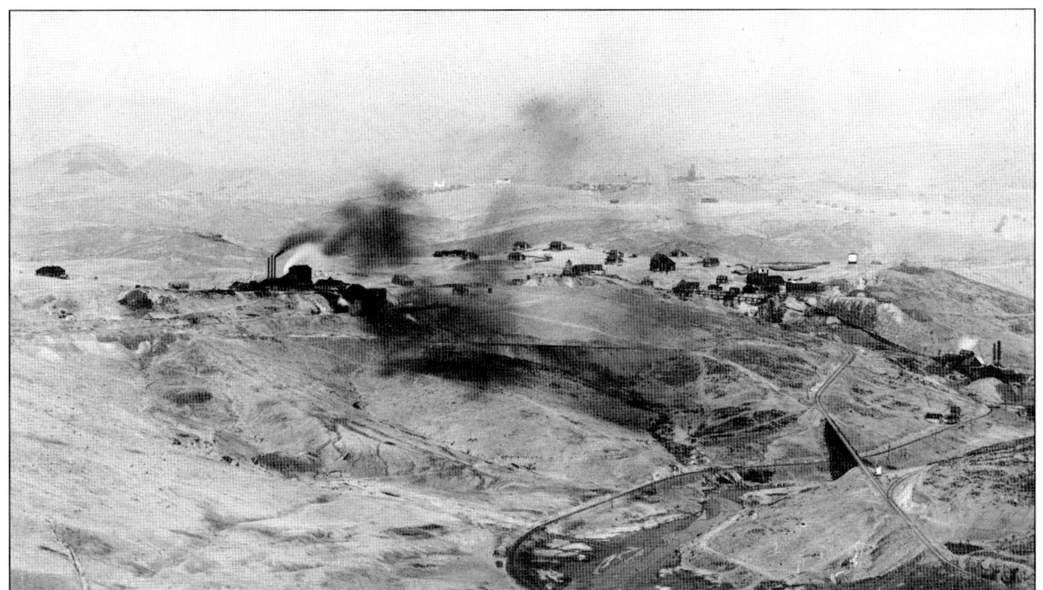

Mining Properties, 1906. The ore of the Ducktown district mines was concentrated in an area of only a few square miles. In this photograph looking north from the smelter chimney, TCC's Polk County Mine (left) adjoins DSC&I's Mary Mine (right). The headframe of TCC's Burra Burra Mine is visible in the distance. (EHW.)

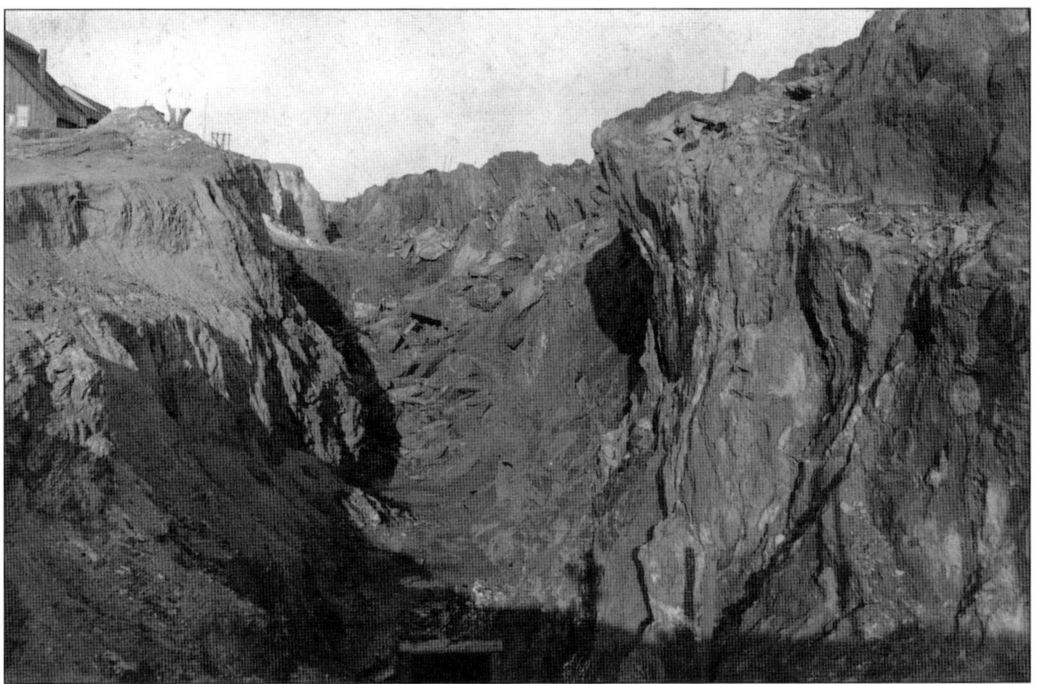

Open Cut Iron Ore Workings, 1905. From 1899 to 1908, the Virginia Iron, Coal & Coke Company mined gossan iron ore in the Ducktown district, paying royalties to TCC, DSC&I, and the school commission. Neither TCC nor DSC&I ever actively mined iron, but iron concentrates from the raw ore were later recovered for commercial use through the two companies' flotation mill processes. (EHW.)

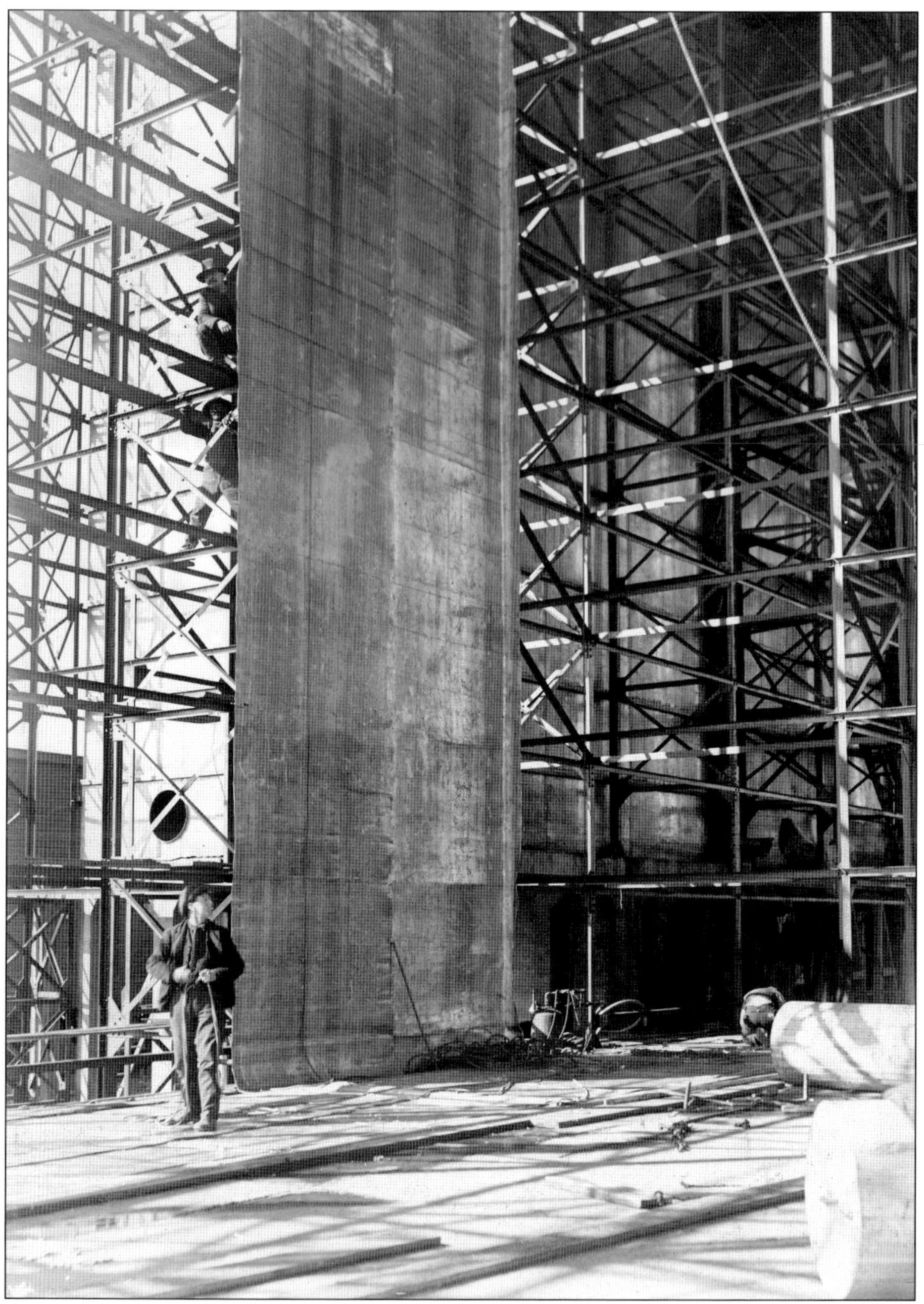

Sulfuric Acid Plant, c. 1907. As the US Supreme Court pondered its decision on smoke damage lawsuits filed by the State of Georgia against both TCC and DSC&I, the two companies separately made the decision to move forward with plans for the construction of sulfuric acid plants to mitigate the environmental impacts of their smelting processes. In 1906, TCC began construction of a huge lead chamber acid plant at Copperhill to convert the noxious smelter gases into marketable sulfuric acid. The plant was completed in November 1907 and shipped out its first tank cars of acid two months later, in January 1908. DSC&I followed suit, opening its own acid plant in June 1909. (EHW.)

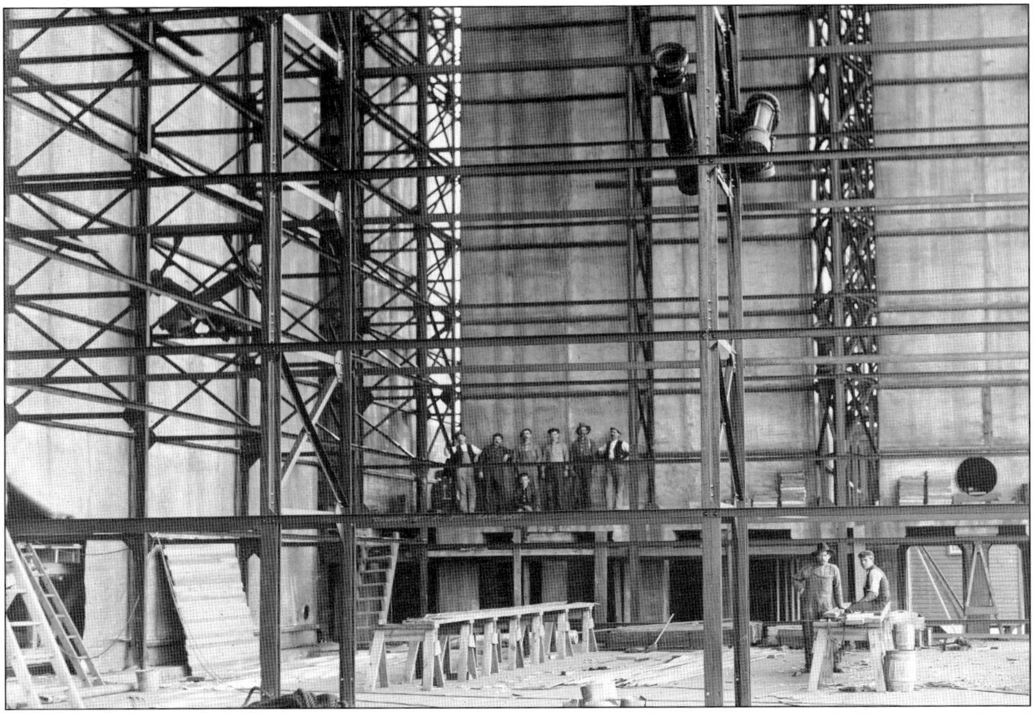

ACID PLANT CONSTRUCTION, C. 1907. When the TCC acid plant at Copperhill was completed in late 1907, it became the largest lead chamber acid plant in the world—and the first ever to operate from copper blast-furnace gases. The structure included 12 lead-lined Falding high chambers, each 70 feet high, along with two Glover towers and four Gay-Lussac towers that were employed at the beginning and end stages of the complex lead chamber process. The new plant was capable of producing 120,000 tons of sulfuric acid a year. (Both, EHW.)

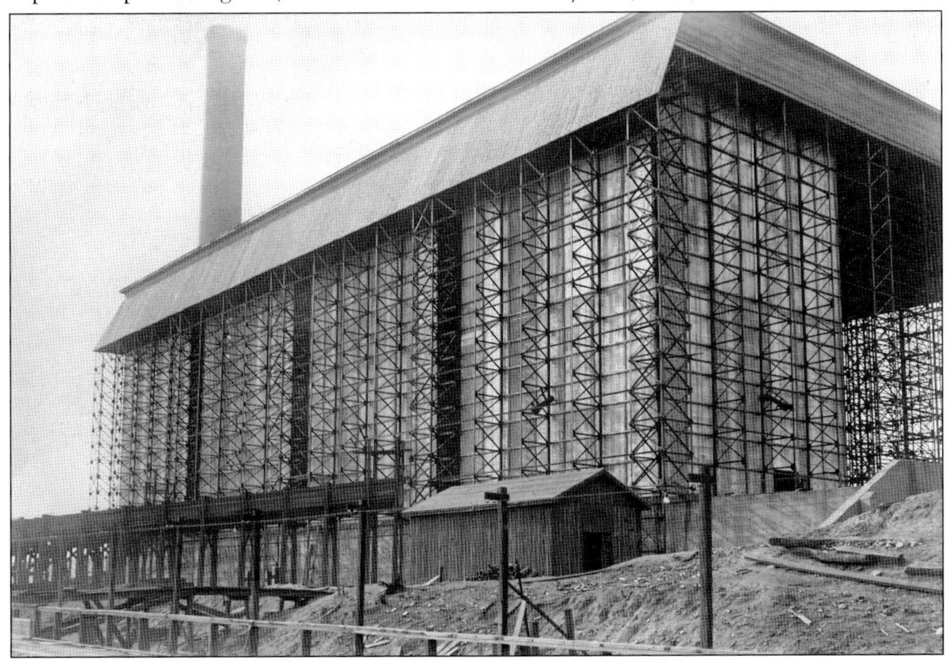

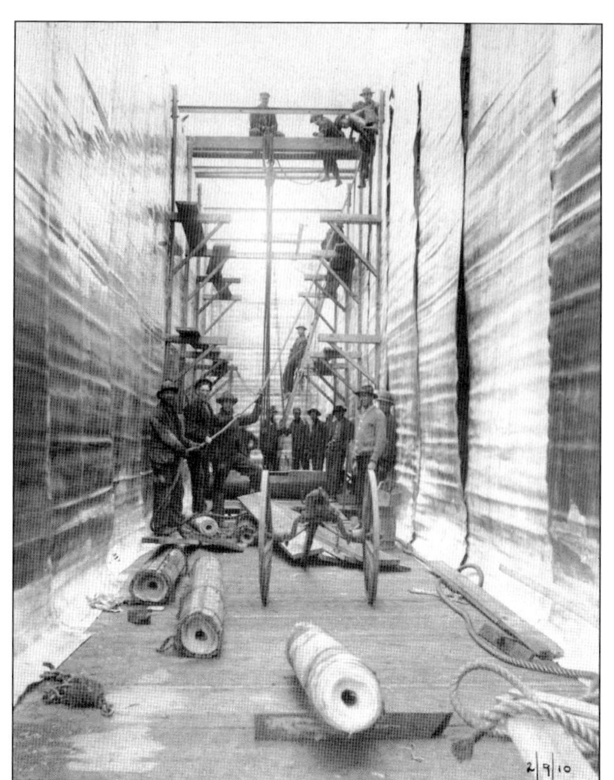

Acid Plant Expansion, 1910. To the surprise of company executives, who had originally conceived it primarily as a means of dealing with smoke damage litigation resulting from the escaping sulfur fumes, the sulfuric acid plant became an unexpected source of profit for TCC. Market demand quickly exceeded supply, and the company began construction in August 1909 on an expansion that would double the size of the original plant.

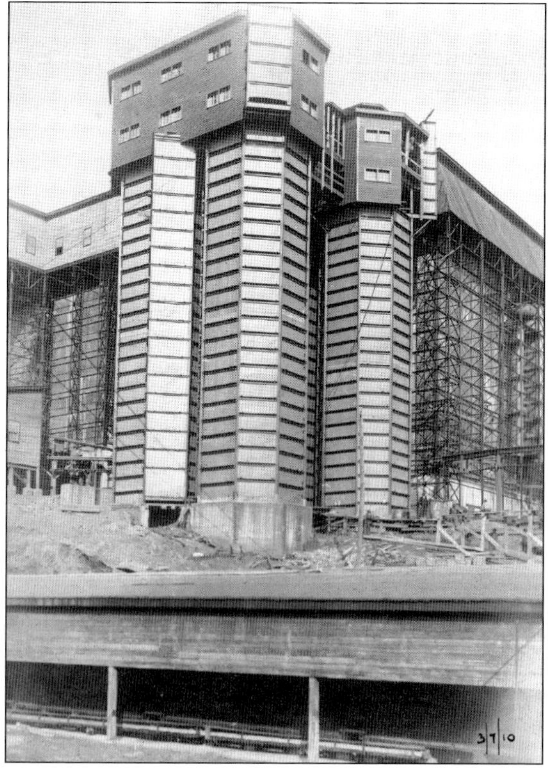

Acid Plant Exterior, 1910. Construction is still underway in this March 1910 photograph. The new expansion started partial operation in late June of that year, even as additional chambers were being constructed. By early December, the expanded plant was in full operation, with an annual capacity of about 225,000 tons of acid. The plant facility required a workforce of only 30 men.

SLAG POURING, EARLY 1900S. The molten black slag from the blast furnaces was tapped into pots (right) that ran along tracks to the high bank in front of the plant, where the red-hot contents were poured down the side of the slope. Below, in 1906, the slag dump can be seen spreading across a wide expanse of the bank above the railroad yard. As of 2021, the giant bank of slag was still being broken up and cleared away by Copperhill Industries, which was founded in 2011 with the primary goal of recycling the millions of tons of by-products resulting from the operations of TCC and its successor companies at the Copperhill plant. (Below, EHW.)

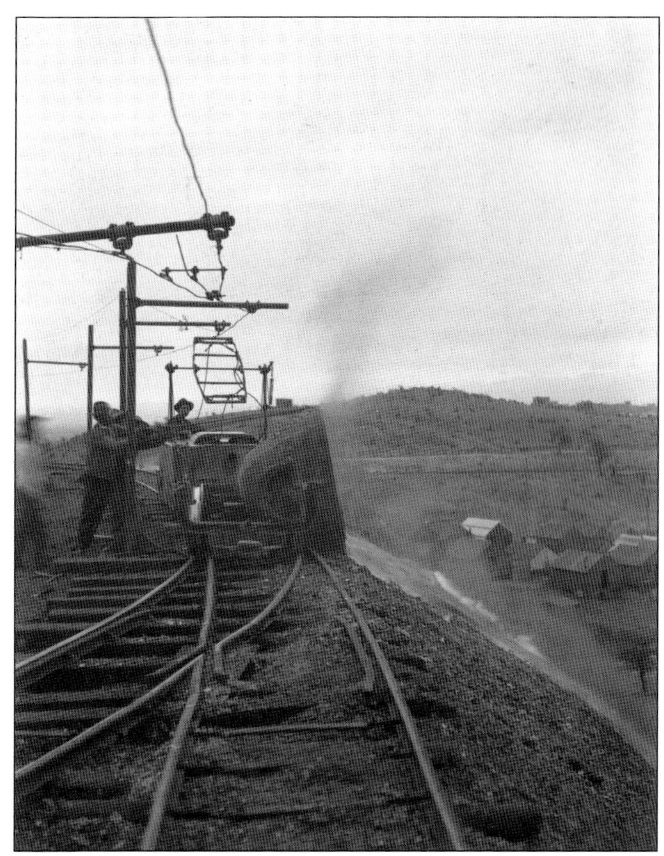

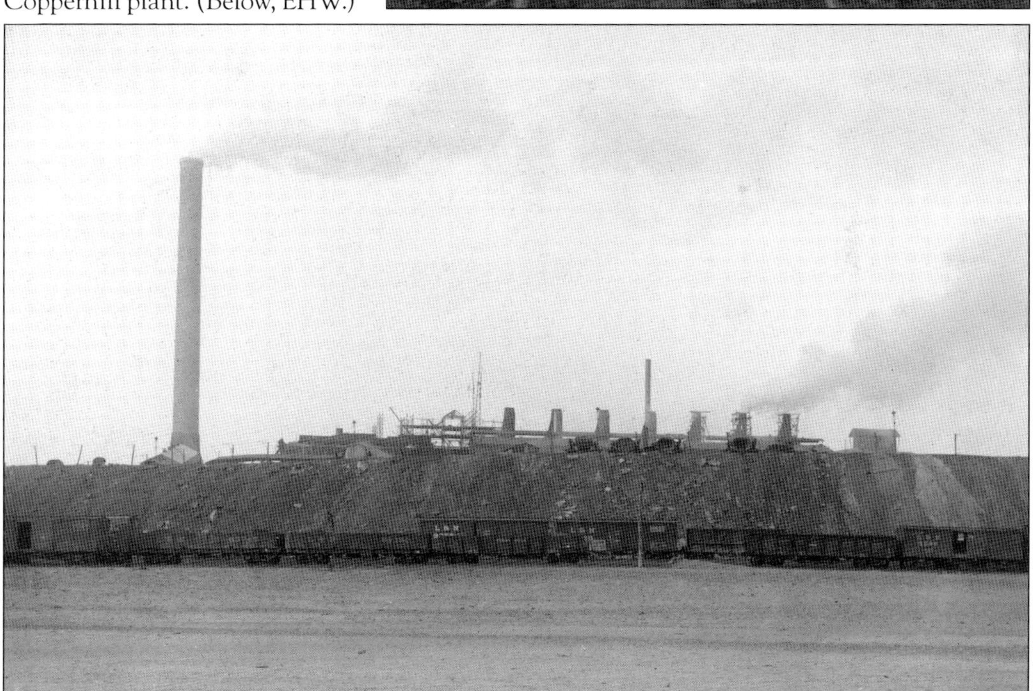

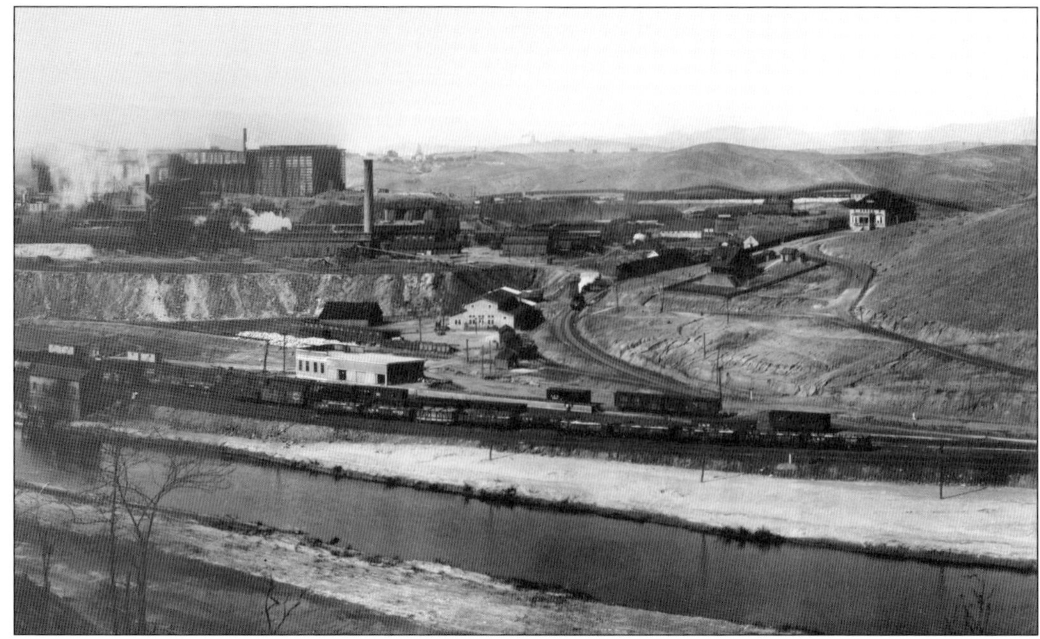

Copperhill Plant, 1911. The newly completed acid plant extension is visible at left to the right of the original plant. The two-level building at center, near the plant entrance, is the Smelter Store, which was built in 1905 after the original commissary on Polk County Hill was destroyed by fire. The hip-roofed General Office is at far right.

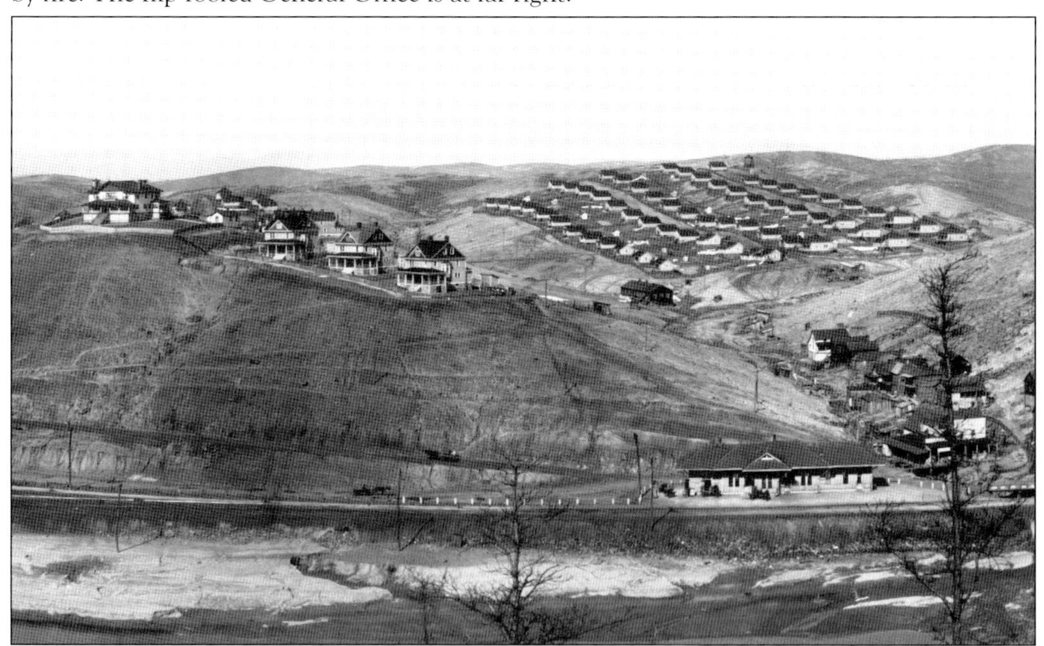

Employee Housing at Copperhill, 1911. By the end of 1904, the company had built about 60 houses for its employees. At Copperhill, that included the rows of small houses known as Newtown, on the right, and larger houses for management, engineering, and other General Office employees on Smelter Hill, at left. At the lower right is the Louisville & Nashville Railroad depot, which was built in 1907.

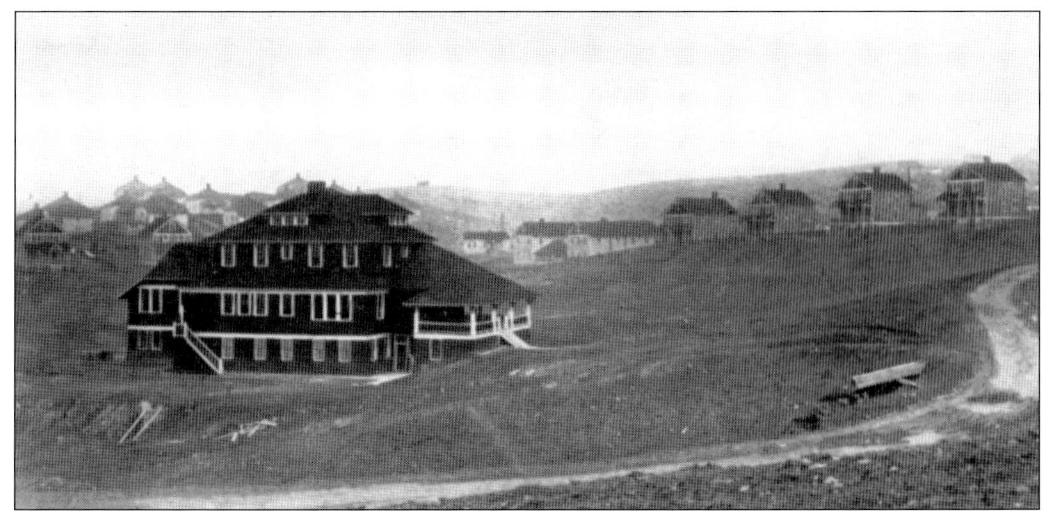

COWANEE CLUB AND SMELTER HILL, 1916. In the early 1900s, members of the TCC engineering staff organized the Cowanee Club, a private club with tennis courts, a swimming pool, a gymnasium, a billiard room, a bowling alley, a kitchen, guest rooms for visiting company executives, and a large main hall that could be used for social events. This portion of a 1916 postcard view shows the club's setting among the houses of the growing Smelter Hill community. The light-colored building between the club and the row of houses at right is the Blue Goose hotel, which primarily housed unmarried employees.

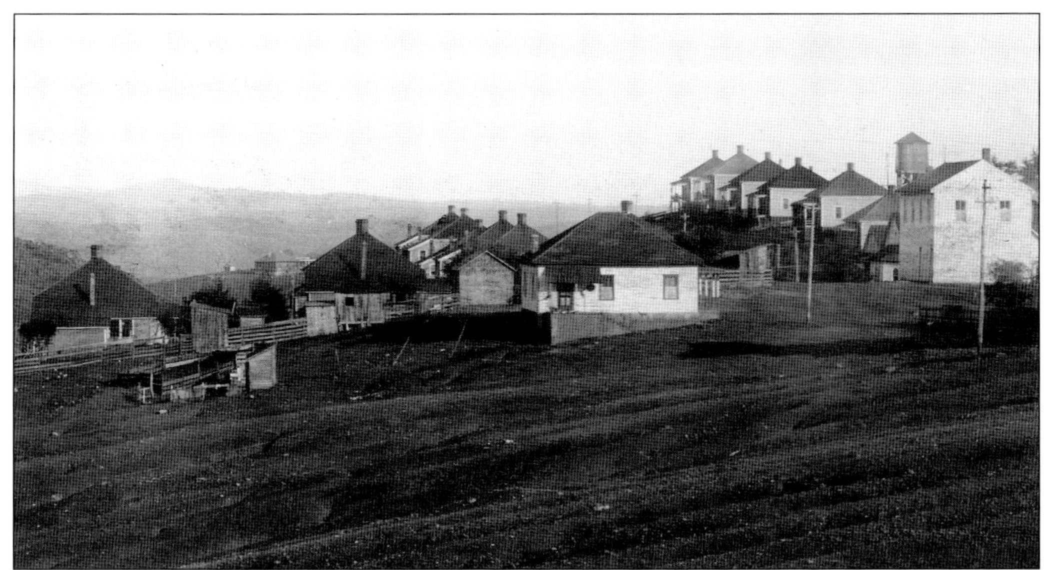

EMPLOYEE HOUSING AT DUCKTOWN, C. 1920. More than 100 copper mine employees were counted in the 1920 census of Ducktown, the site of TCC's large and productive Burra Burra Mine. Although some of them may have been employed at the nearby Isabella workings of DSC&I, most are more likely to have worked at Burra Burra. The small company houses at Ducktown, like those at Copperhill, were virtually identical.

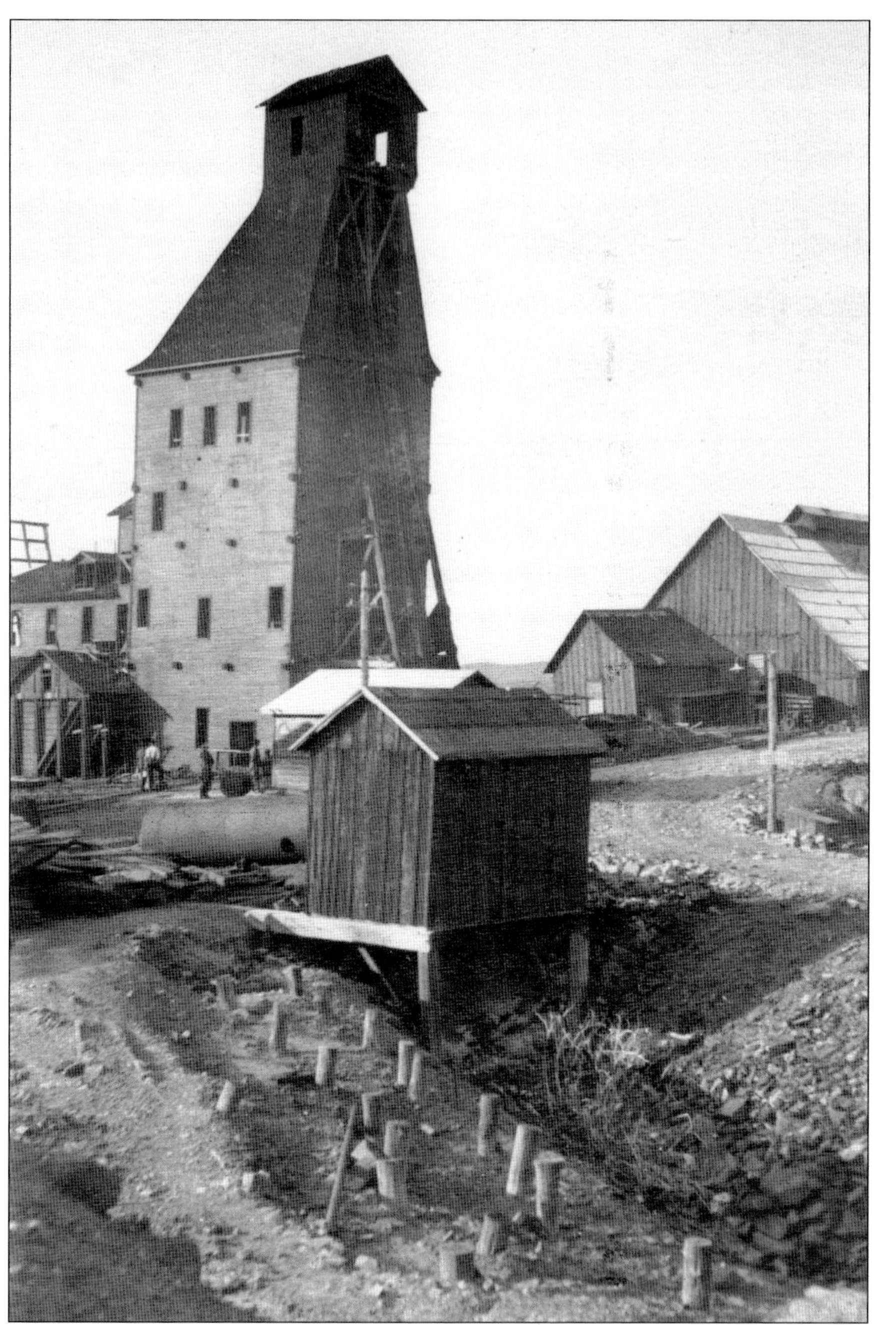

Burra Burra Mine, Early 1900s. In 1899, armed only with a carpenter's framing square and a shovel, Swedish-born mine captain C.V. "Cap" Skullman marked off the original shaft collar for TCC's Burra Burra Mine. He is also credited with lifting the first shovelful of dirt from the shaft. During its nearly 60 years of operation, Burra Burra yielded more than 15 million tons of ore. "The meat and bread for hundreds of families have come out of that hole in the last sixty years," said one veteran miner when the mine was finally closed down in the summer of 1958. Today, the mine complex, which is largely intact except for the headframe and a portion of the ore bins, is the site of the Ducktown Basin Museum and is listed in the National Register of Historic Places.

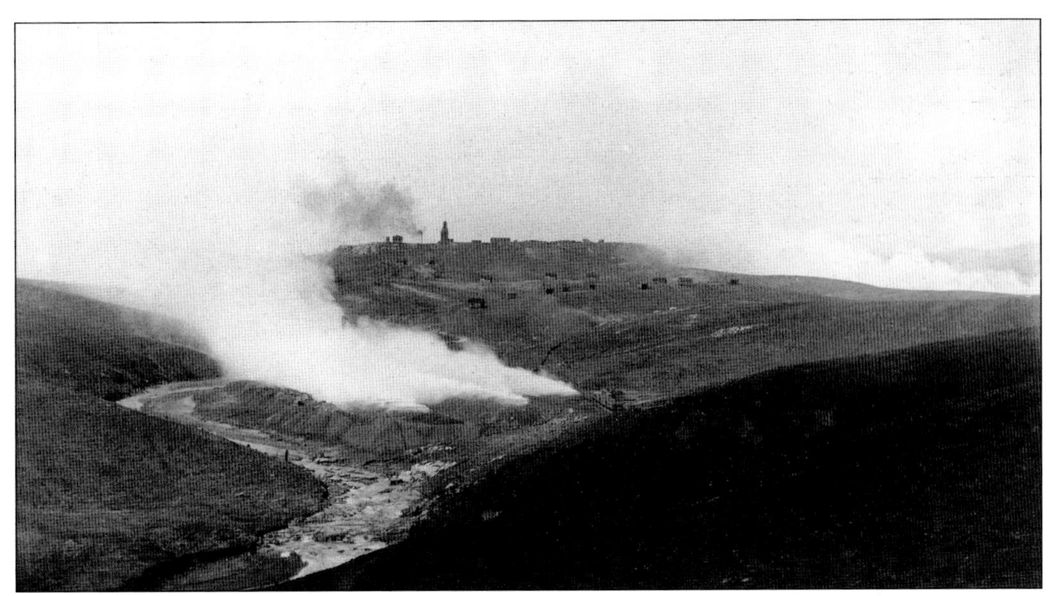

Burra Burra Vicinity, Early 1900s. In the 1903 photograph above, smoke rises from the newly constructed Buena Vista roast yard at center and the 14-acre Burra Burra roast yard at right. The Burra Burra Mine complex is at top center, above the level of the roast yards, with smoke emerging from the chimney of the boiler house. The preliminary roasting of the ore in open sheds, which had been the standard method in the 19th century, produced a dense sulfurous smoke that had already destroyed the surrounding vegetation long before the advent of TCC. Below, looking south from the Burra Burra site in 1905, the barren landscape stands out even more starkly against the scattered buildings. The smelter at McCays is visible on the horizon. (Both, EHW.)

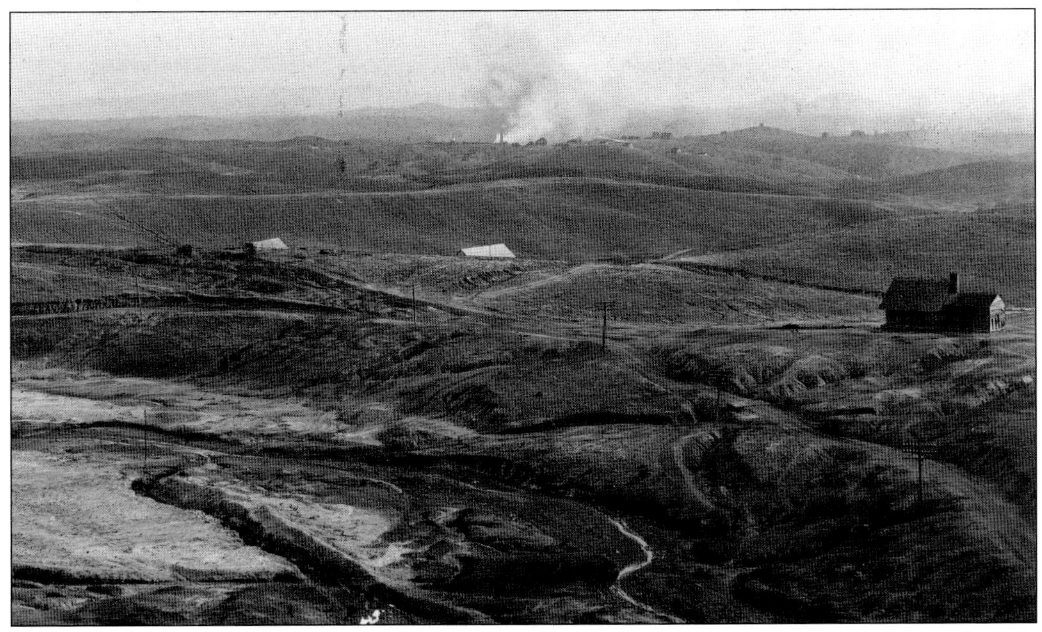

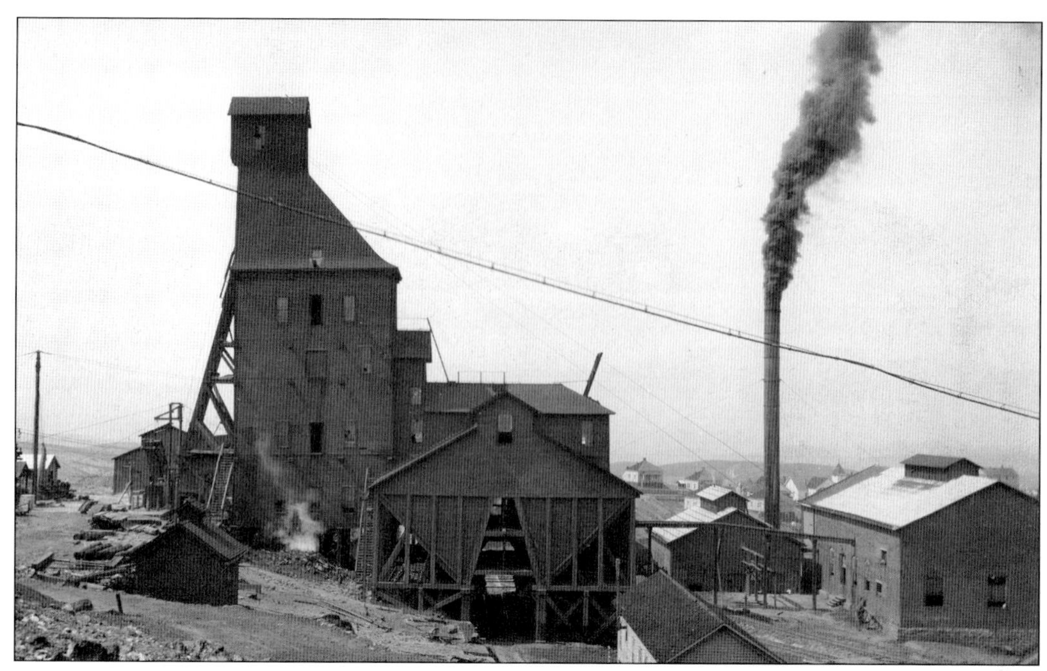

STEAM POWER AT BURRA BURRA, EARLY 1900S. Like the other mines in the Ducktown district, Burra Burra was originally powered by steam. At right above, smoke emerges from the boiler house, which supplied power to the hoist engine and air compressor in the adjacent hoist house (far right). The boiler house also supplied power to the other buildings on the site. Small company houses in Ducktown are visible in the background. The pre-1910 photograph below, taken from a different angle, includes some of the smaller structures on the site. Original construction at Burra Burra was reported to be largely complete by 1901.

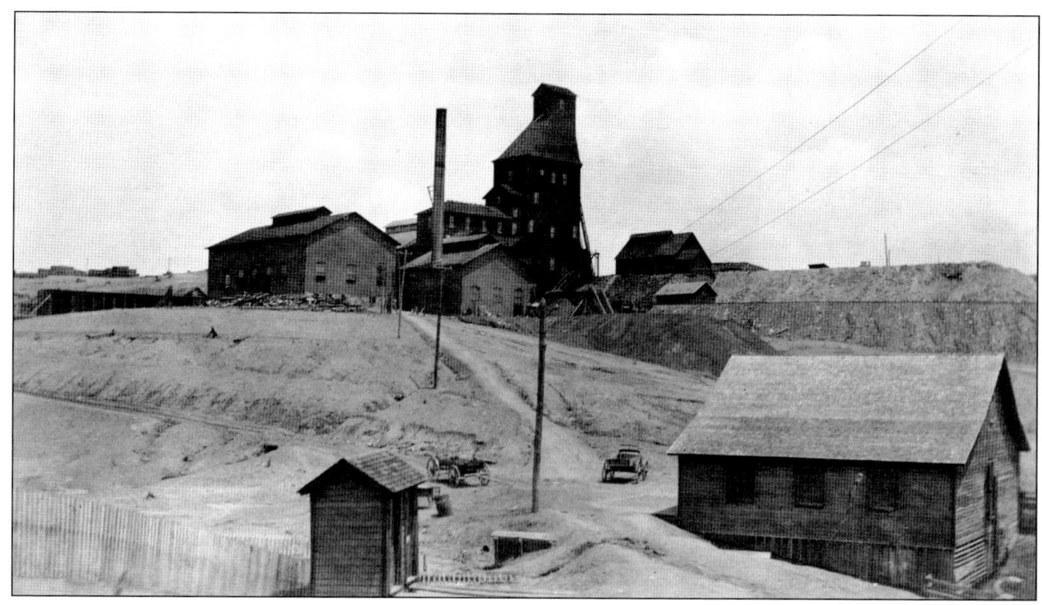

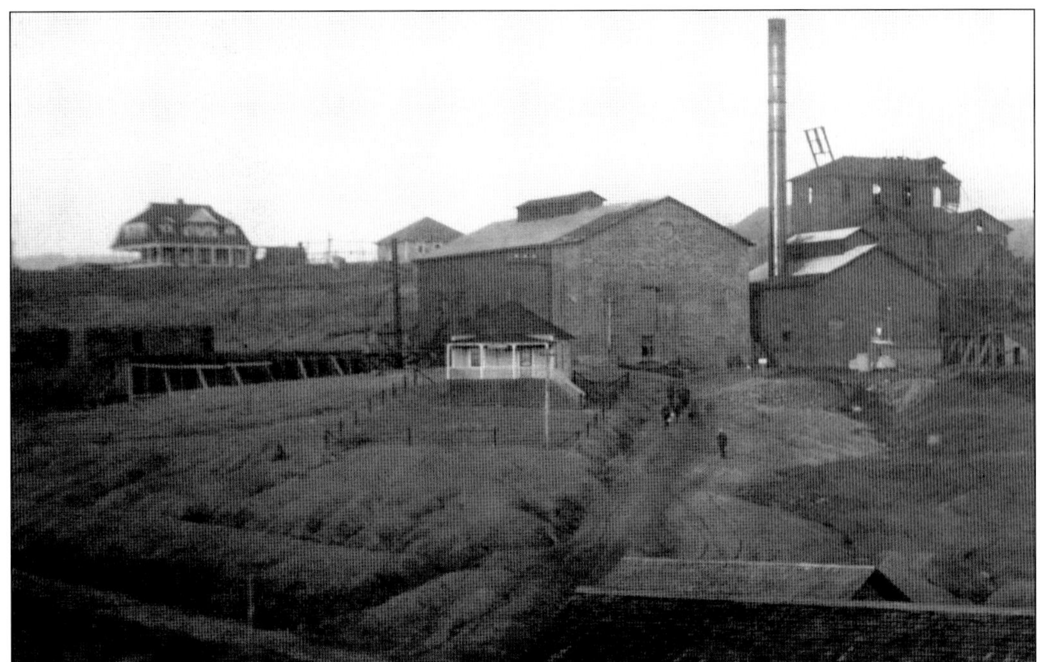

Construction at Burra Burra. In summer 1916 (above), the wooden headframe and ore bins at Burra Burra were demolished. In summer 1917 (below), a new steel headframe and concrete ore bins were erected in their place. While construction was underway, ore was hoisted at McPherson shaft, which was connected to Burra Burra by an underground drift. During rebuilding, both Burra Burra and McPherson were switched from steam power to electric power from a nearby Ocoee River plant. The small white building at center above is the employee clinic, built in 1910, and the larger building at left is the staff house residence for mining department employees.

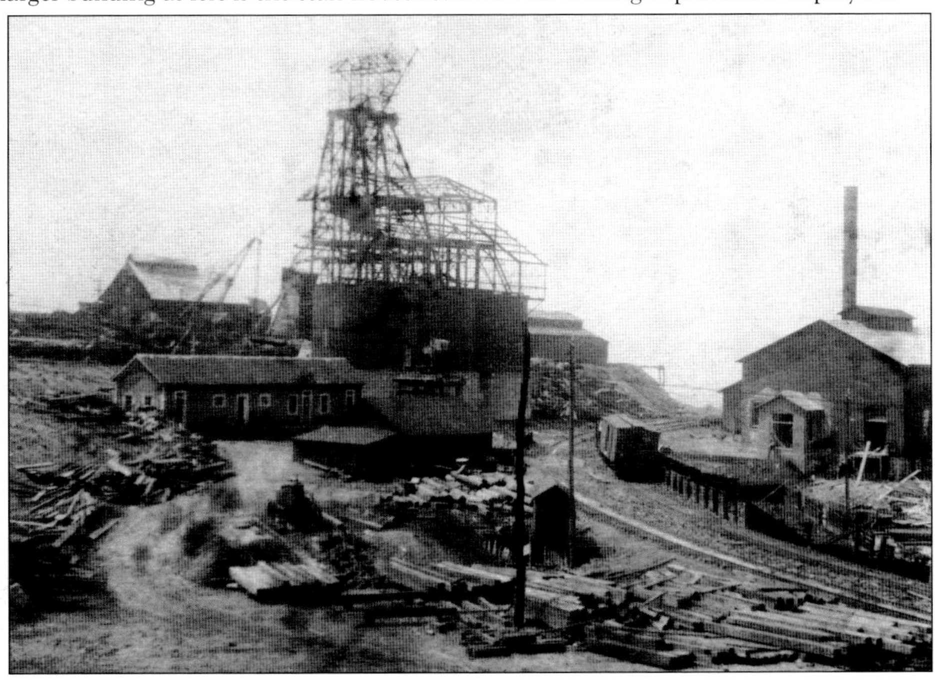

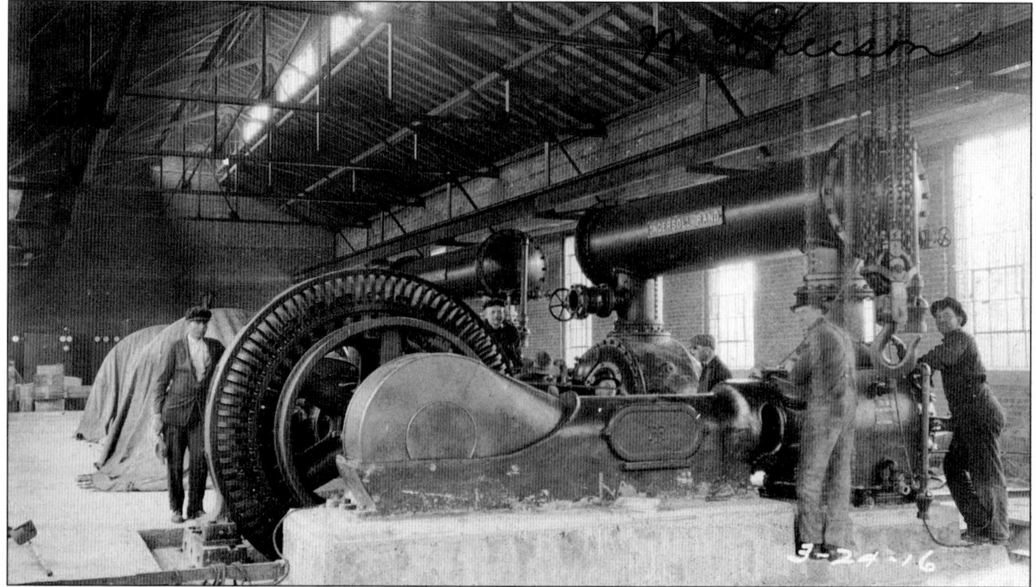

Electric Hoisting Engine at McPherson, 1916. Exploration of the McPherson ore body began in 1905, but the McPherson shaft was connected to the Burra Burra workings by an underground drift in 1908 and was subsequently considered part of Burra Burra Mine. In early 1916, a new compressor house, crusher house, and powerhouse were built at McPherson, and electric power replaced steam power. (EHW.)

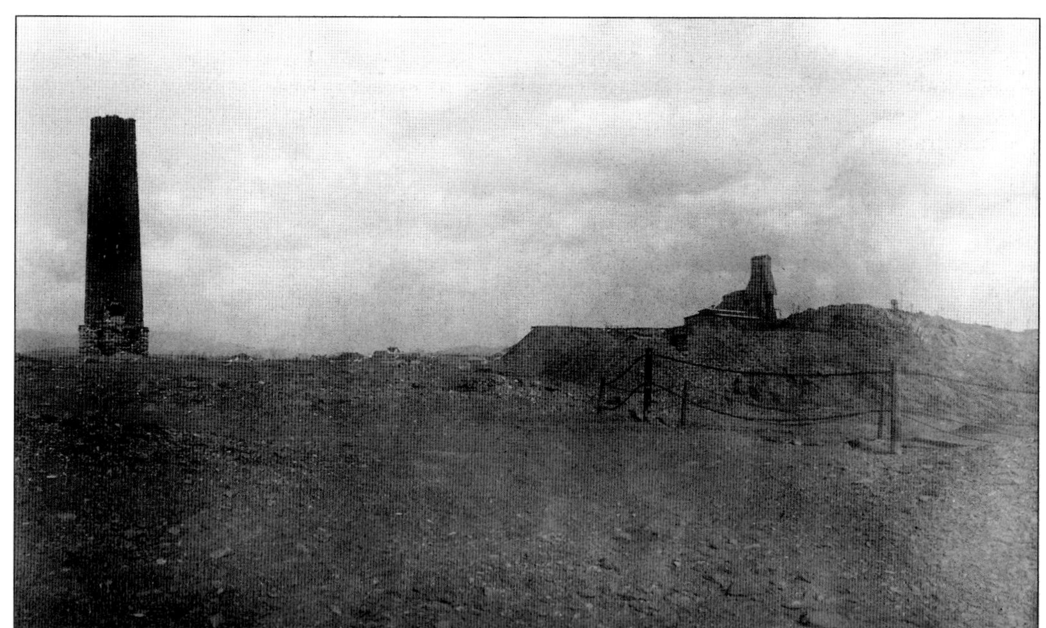

Burra Burra and Hiwassee, c. 1920. In 1908, TCC also drove a separate underground drift connecting Burra Burra to the old Hiwassee Mine, effectively opening up a single ore body that could be explored along its entire length, from McPherson to Burra Burra to Hiwassee. In this photograph, the old Hiwassee Mine chimney stands in the foreground at left, and the Burra Burra complex stands on the hill at right.

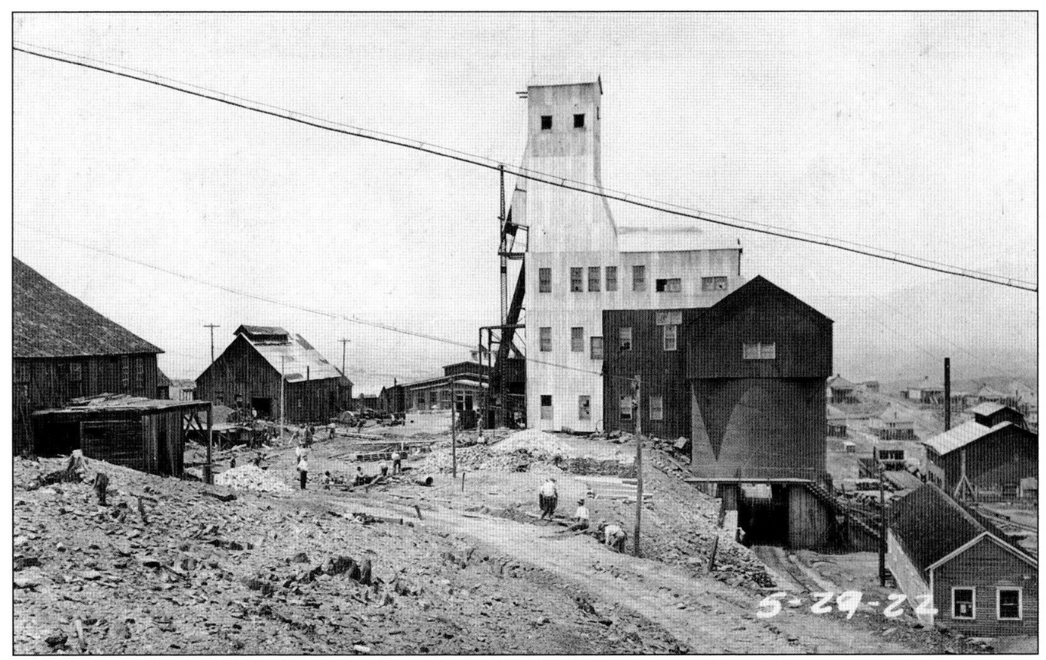

New Shops at Burra Burra, 1922. In the spring and summer of 1922, a new brick shop building was erected at Burra Burra to house the growing number of skilled craftsmen employed by the mine. These photographs, taken just three weeks apart, show the rapid progress of the new building, from ground clearing and foundation construction (above) to near completion of the brick structure (below). Much of the equipment needed in the mining operations was fabricated on-site in the shops, which were also responsible for repair work, painting, and other maintenance tasks. The new change house, built in 1918, can be seen in the center background to the left of the headframe. (Both, EHW.)

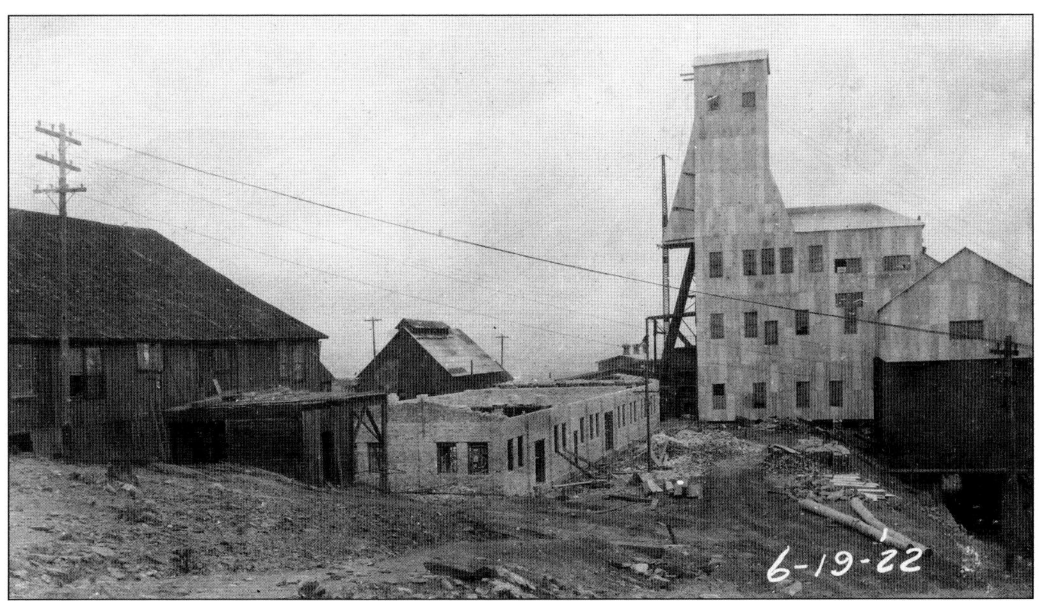

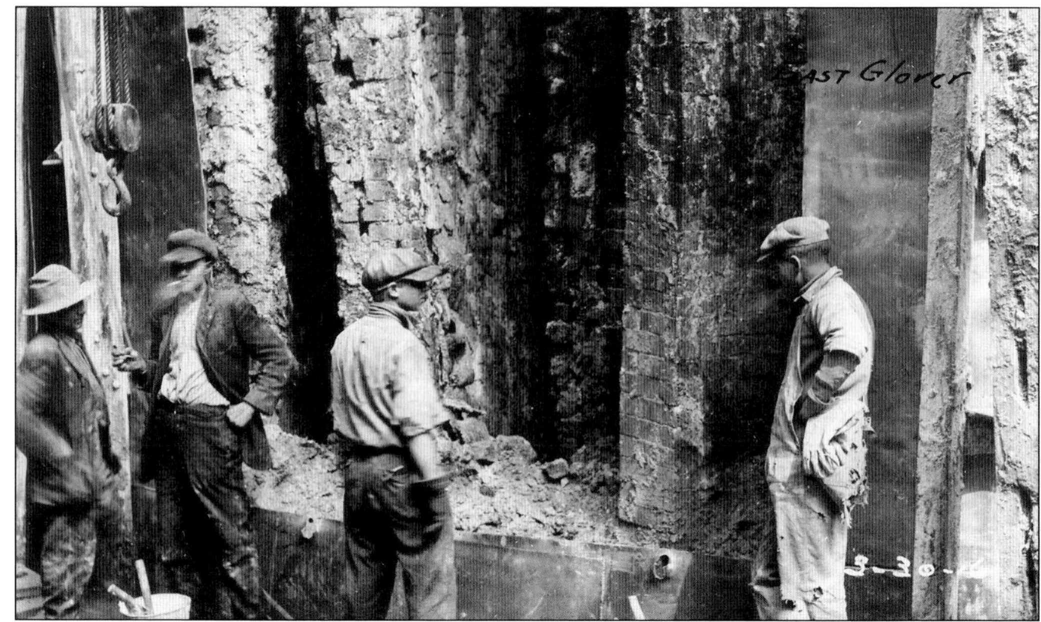

New Acid Plant Construction, 1916. The increased wartime demand for sulfuric acid necessitated the building of a second chamber acid plant at Copperhill. TCC had already negotiated profitable contracts with several companies, including the DuPont Powder Company, which needed a steady supply of sulfuric acid for the production of gunpowder. Complicating matters was a two-month shutdown in early 1916 of the No. 1 acid plant, which had been pushed beyond its capacity and needed to be taken offline for major maintenance. Above, workers on the site of Glover tower construction take a break; below, foundation work on one of the new plant's Gay-Lussac towers illustrates how much manual labor was involved in the project. (Both, EHW.)

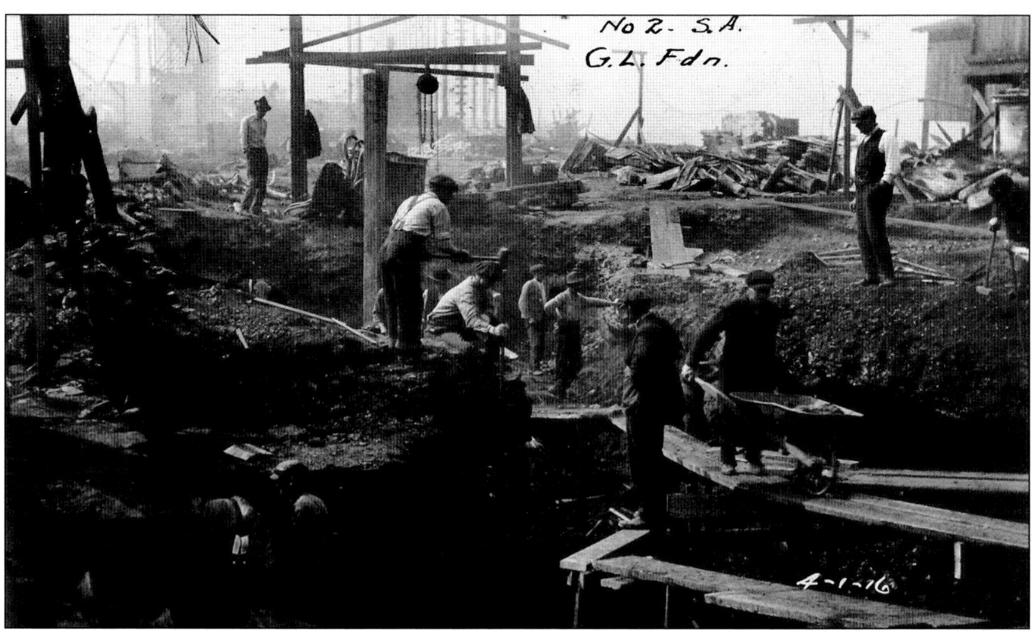

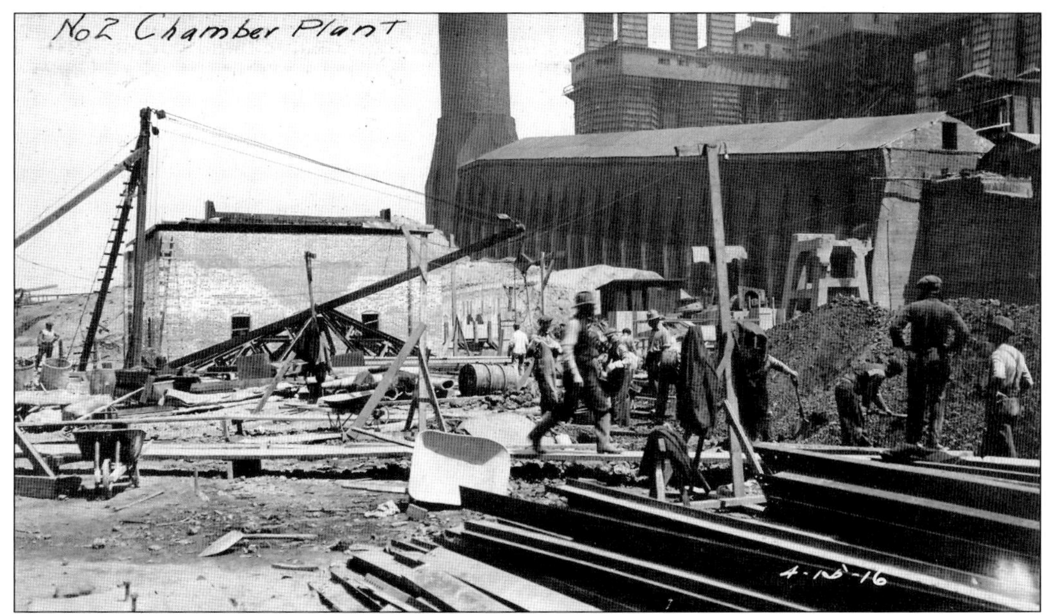

Acid Plant Progress, 1916. The existing No. 1 acid plant can be seen at top right above, separated from the new construction site by the smelter flue and chimney. Below, the skeletal structure in progress shows the long, narrow configuration of the new plant, which was being built to the west of the existing plant. The new No. 2 plant was completed in July 1916 and was started up just two days later using a chamber control process patented by a TCC chemist. At this time, TCC and DSC&I were still the only two companies in the world making sulfuric acid from copper blast-furnace gases. With the addition of the new plant, TCC increased its already prodigious production capacity to 325,000 tons of sulfuric acid a year. (Both, EHW.)

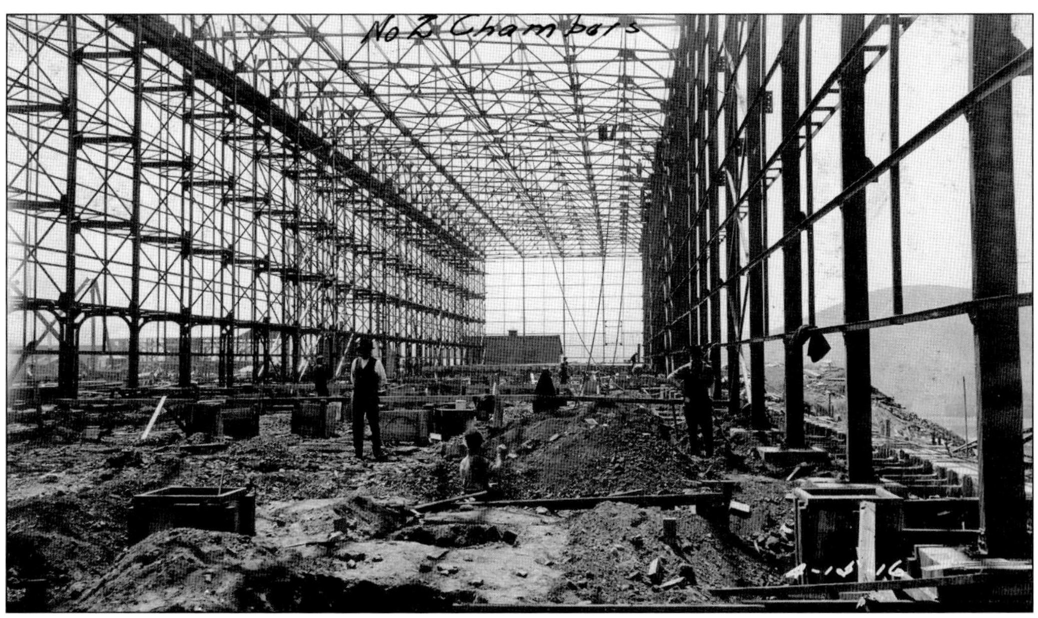

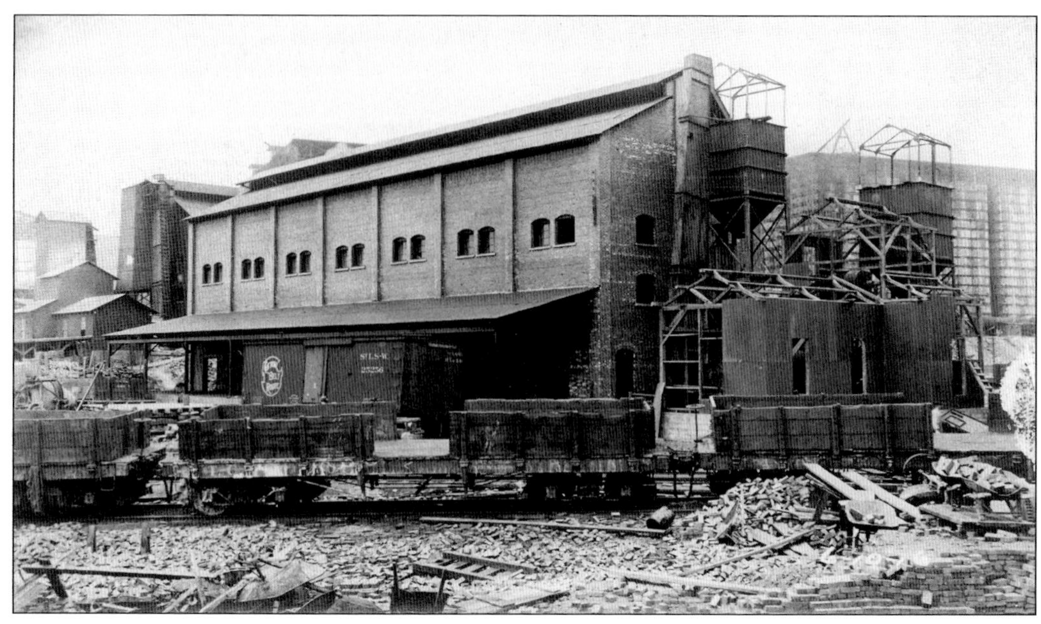

New Acid Facilities, 1915–1916. As wartime production demands increased, so did the complexity of TCC's acid operations. A new nitric acid plant (above), built in 1915, provided an added source of income for the company as well as providing nitric acid to be used in the sulfuric acid process. Two more nitric acid plants were built the following year. Also put into wartime operation to provide for different grades and strengths of sulfuric acid were a reverberatory acid concentrator (below), a grade B plant, several oil of vitriol plants, and a Cottrell precipitator plant for treating sulfuric acid mist. Some of the plants did not operate successfully and were soon removed from operation. (Below, EHW.)

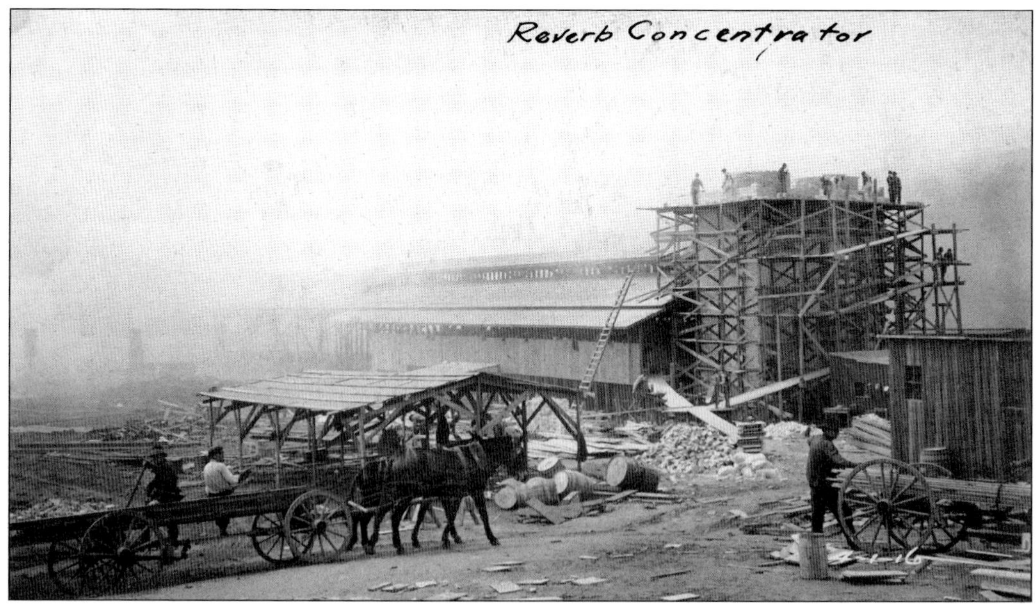

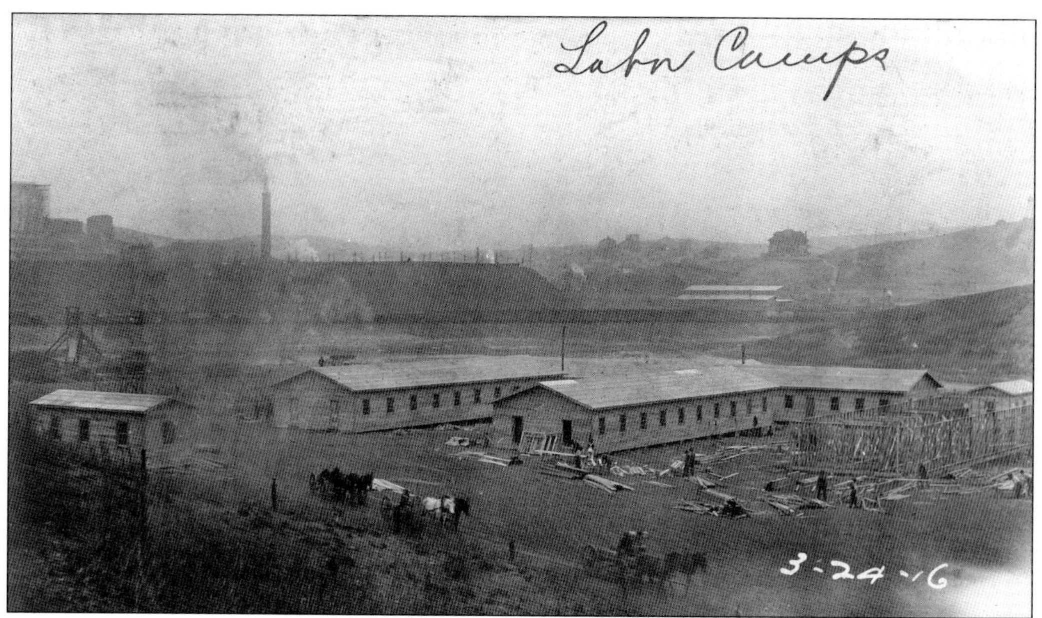

New Company Housing, 1916. Increased wartime production also required increased housing for the influx of employees needed to operate the new facilities at TCC's Copperhill plant. In early 1916, the company began constructing additional dormitory-style housing in its existing foreign workers' camp (above) just across the river from the plant. On Smelter Hill (below), on the other side of the river, the company not only built additional houses for its office and engineering staff, but also equipped the new residences with indoor plumbing, which was apparently worth noting on the photograph. During the same period, workers' housing at Newtown was extended to include new houses on the adjacent slopes of Cemetery Hill. (Both, EHW.)

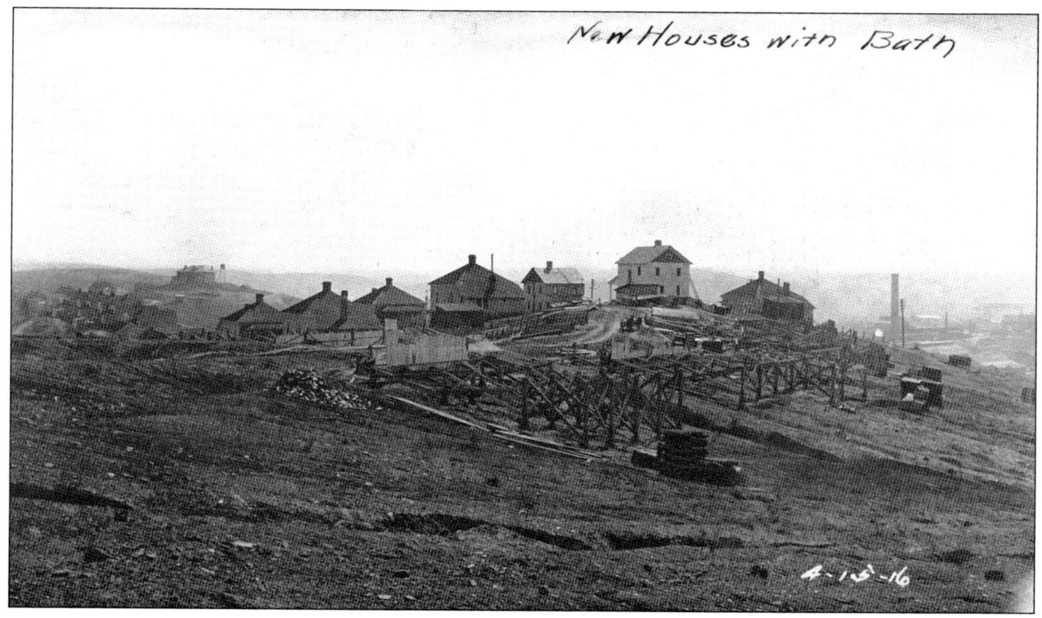

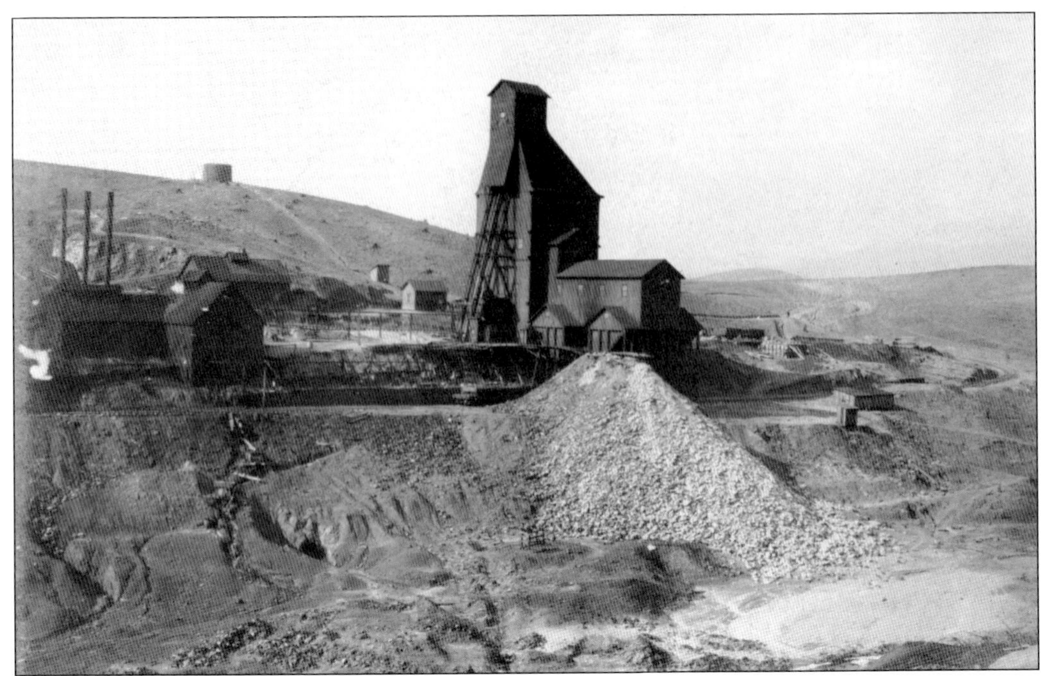

London Mine, c. 1904. London Mine, which first opened in 1853, was reopened by TCC in 1901 and continued in operation for 25 years. This photograph was taken around 1904 by Edward Oscar Boak. The mine was closed permanently in 1926, a few years after construction of the London Mill flotation plant at the site.

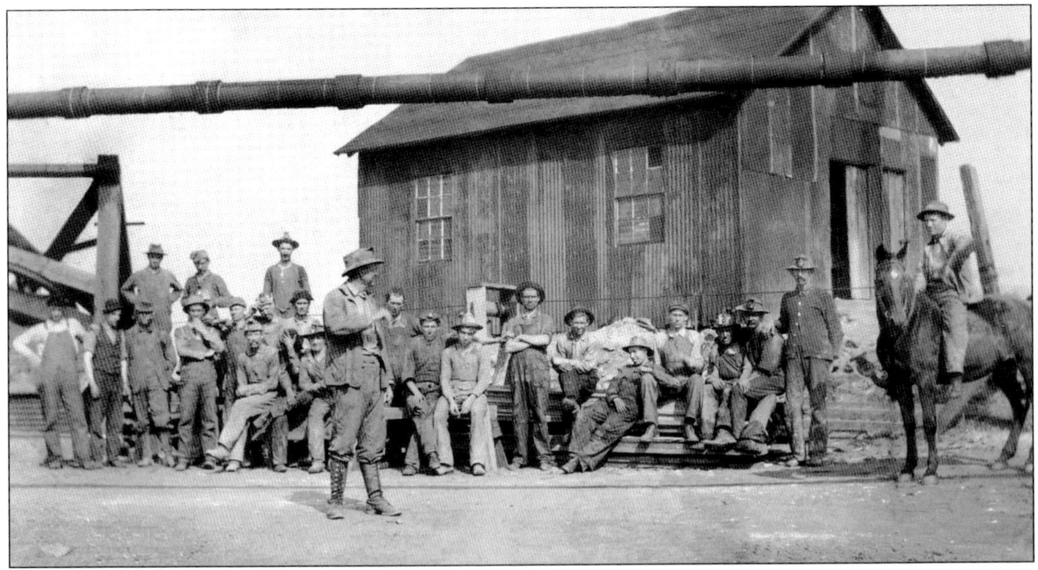

London Mine, Early 1900s. This undated photograph at London Mine includes several workers wearing the oil-burning headlamps known locally as "coon-shiners" due to their use of raccoon oil as fuel. TCC began the switch to carbide headlamps between 1910 and 1912, so this photograph seems to have been taken prior to the changeover. Seated fourth from right is Lucius Burger. (Courtesy of Thomas E. Panter III.)

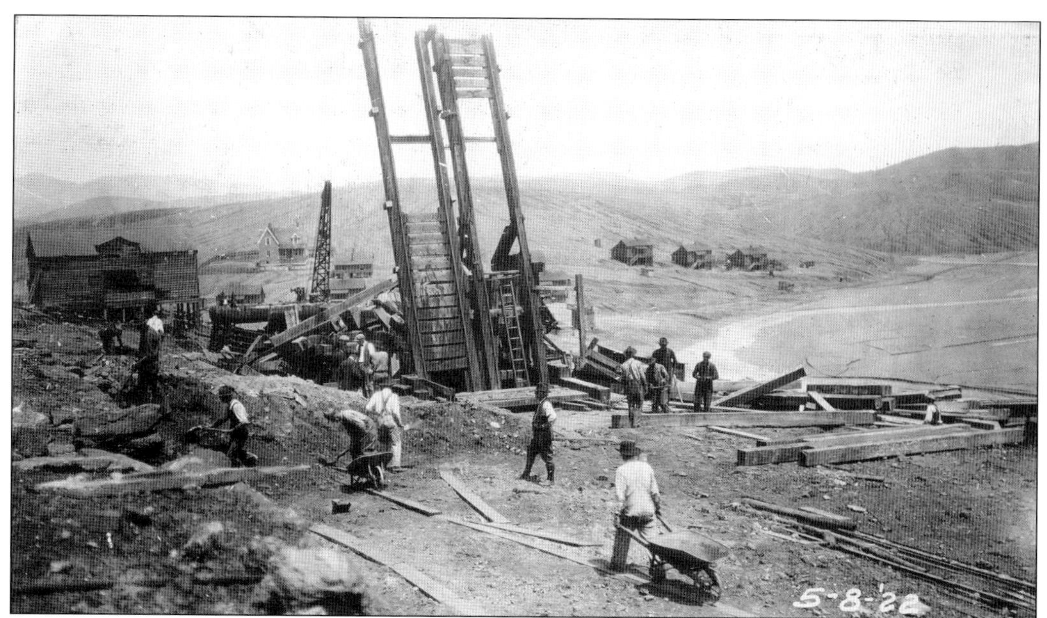

London Mine and Mill. In 1920, TCC began planning for a flotation mill at London Mine with the objective of eliminating the blast furnaces and increasing copper production from lower-grade ores. Construction, which also involved demolishing and rebuilding the surface structures of the mine, was begun in 1922 (above). In addition to increasing copper production, the changeover allowed for the separation of other minerals, particularly iron and zinc, that could not be successfully recovered during the traditional smelting process. In 1925, new roaster and sinter plants were built at Copperhill to process the recovered iron concentrates. Fourth from left in the photograph of London Mill workers below, which is thought to date from the 1930s, is Lucius Burger. At far right is George Bowers. (Above, EHW; below, courtesy of Thomas E. Panter III.)

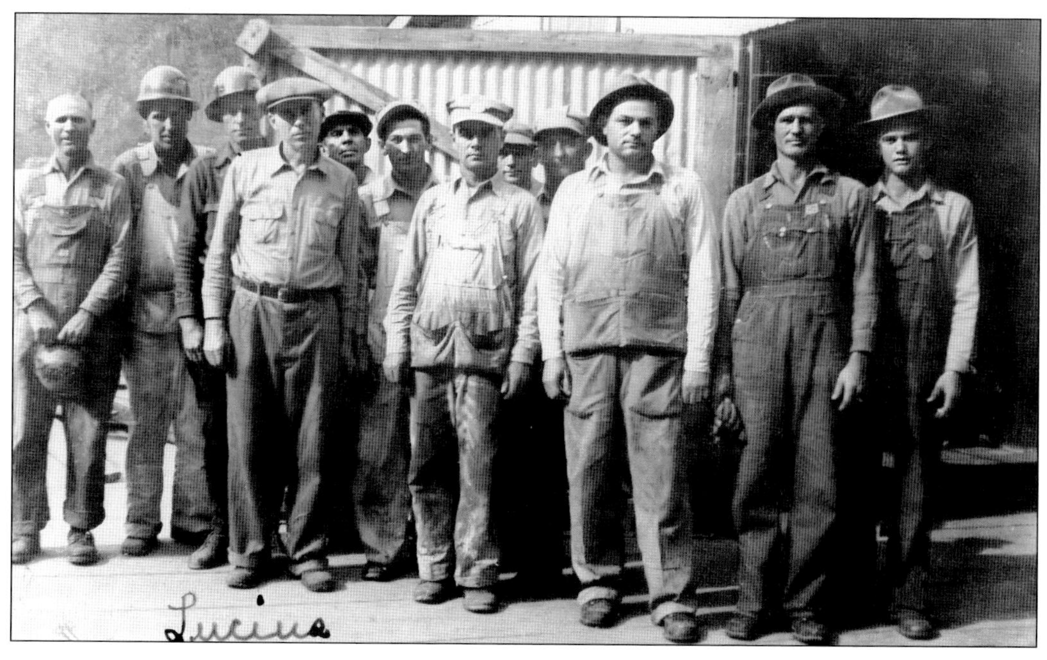

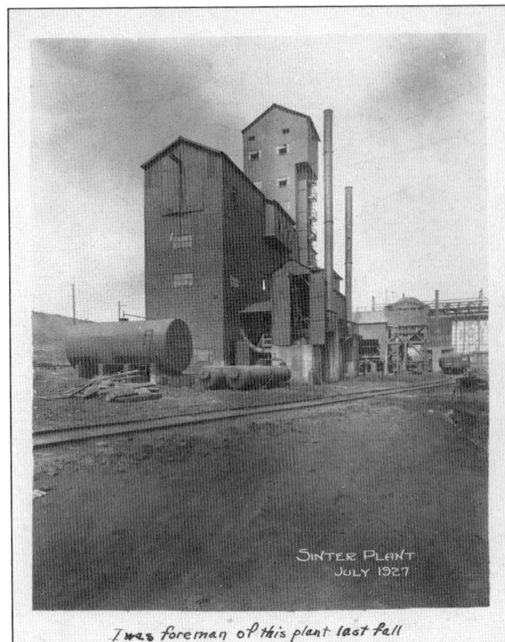

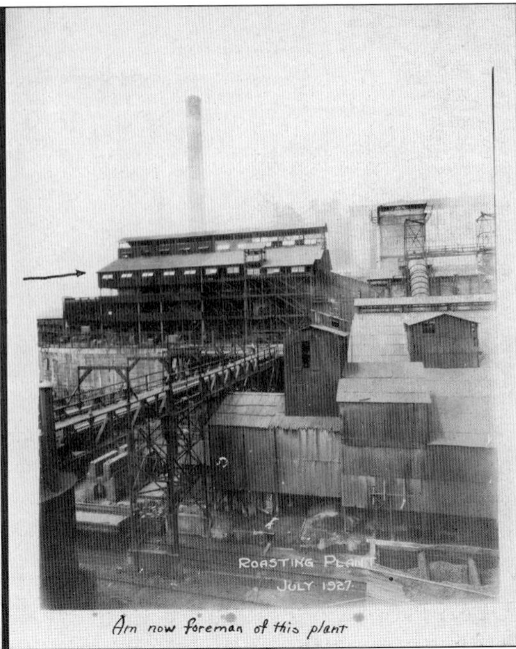

RISING THROUGH THE RANKS. When he arrived at TCC in 1927 as a young chemical engineer, R.R. Burns was assigned to the laboratory at Copperhill. Soon afterward, he became shift foreman at the roaster and sinter plant and then general foreman at the roasters (above, with his notations). A series of promotions followed, culminating in his being named vice president of Tennessee Corporation in 1962. In 1963, Burns succeeded the retiring E.H. Westlake to become Tennessee Corporation's third president. In December 1966, Burns (at right, below) returned to Copperhill to serve as toastmaster at the retirement party of his longtime friend, TCC president and general manager T.A. Mitchell (left). In the background is Burns's wife, Elizabeth.

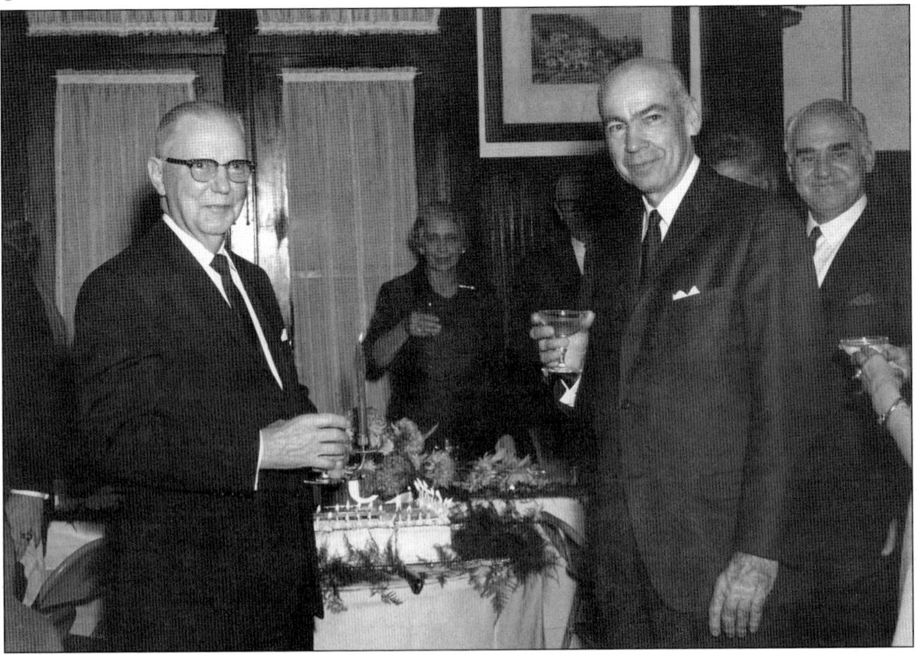

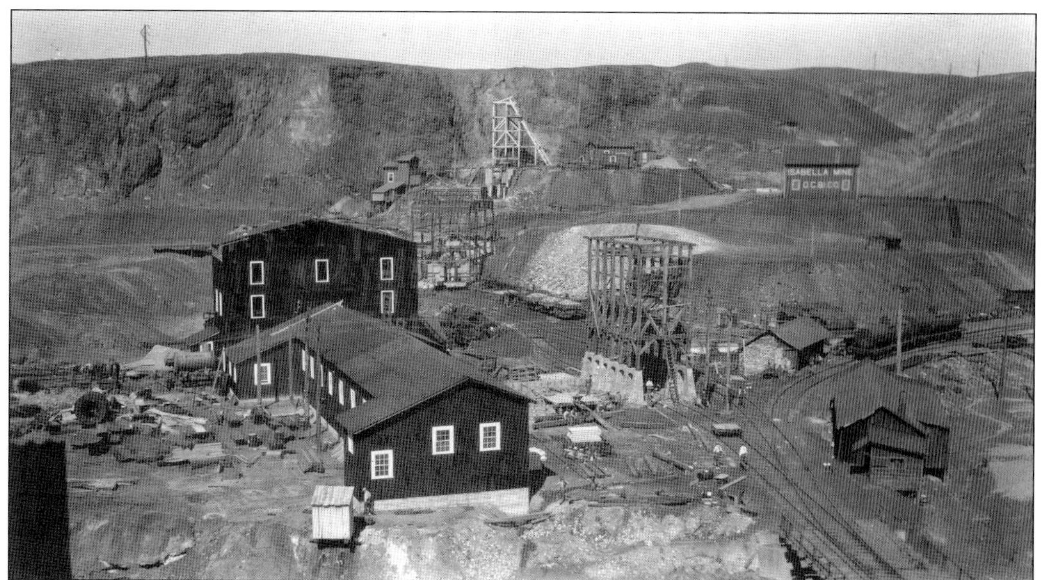

DC&I Properties at Isabella, 1920s. In 1925, after a period of financial difficulties, DSC&I was purchased by the Ducktown Chemical & Iron Company (DC&I). The new company would soon run into financial difficulties of its own and was purchased by TCC in 1936, consolidating all the Ducktown district mining properties under a single management. The photographs of 1927 construction at Isabella (above) and of the Isabella plant in operation (below) are from the files of DC&I general manager W.F. Lamoreaux.

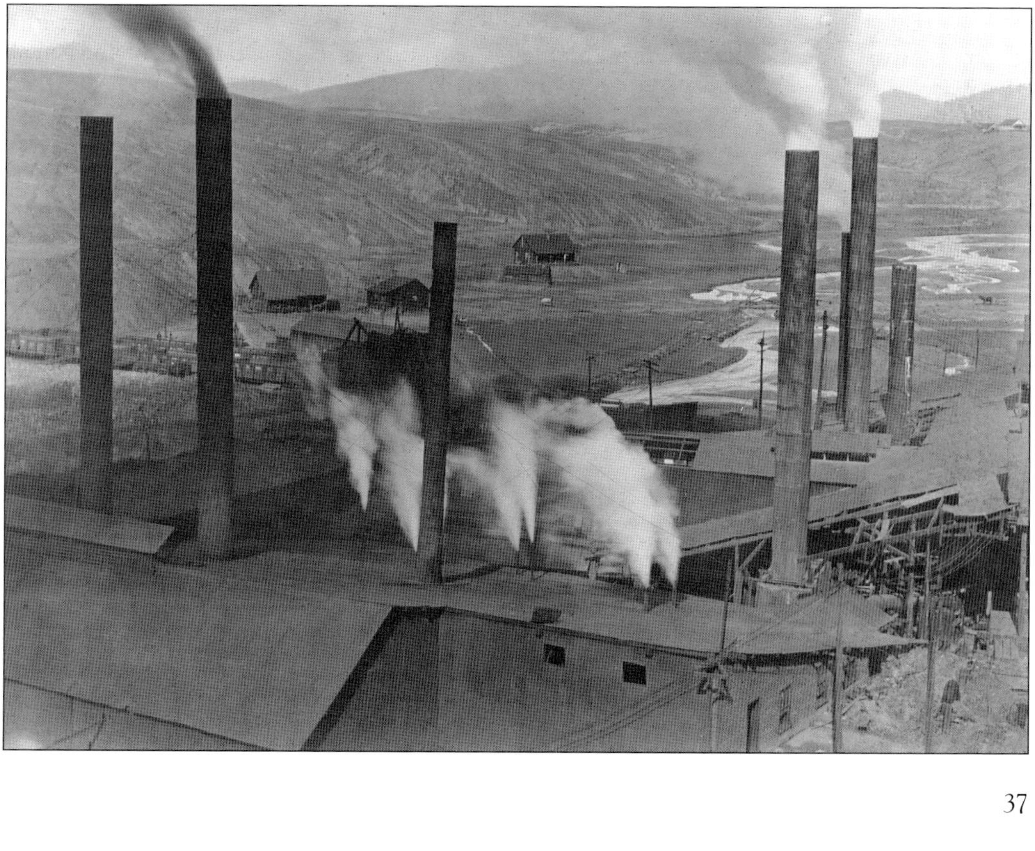

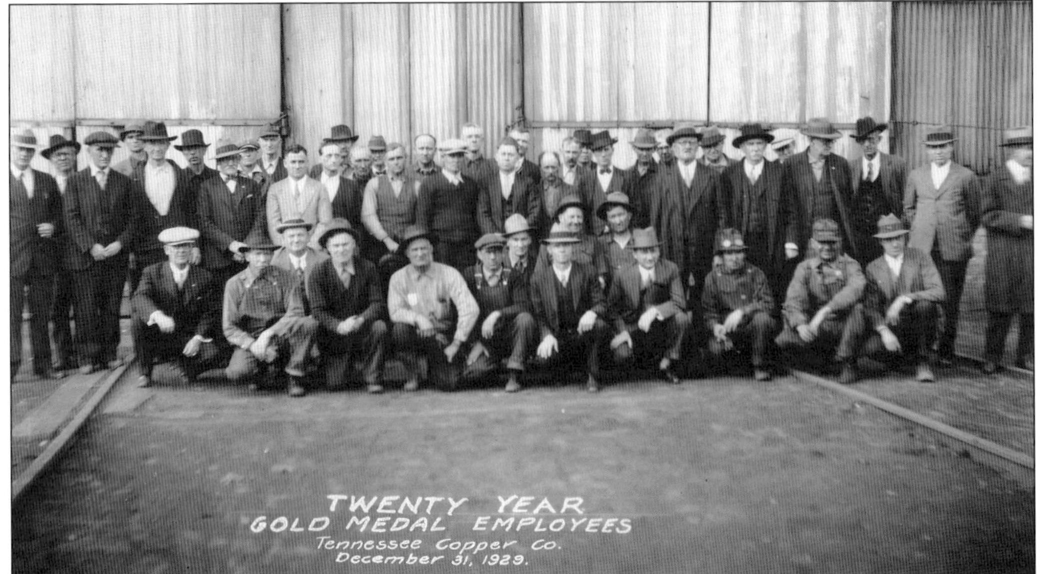

Gold Medal Employees, 1929. TCC tended to retain employees, often for decades. The men in this photograph would have been employed by the company at least since 1909, if not earlier. TCC later acknowledged personal safety anniversaries for employees in its company newsletter, but it is unclear whether the 1929 Gold Medal Employee designation applied to company longevity or to personal safety records.

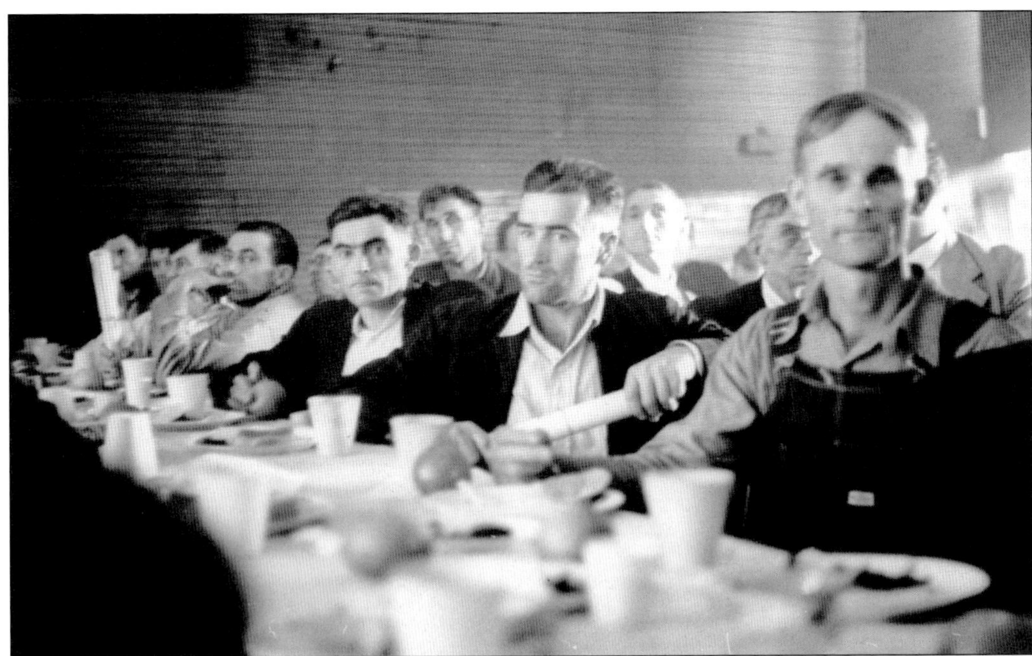

Sentinels of Safety Banquet, 1934. In 1933, the US Bureau of Mines awarded its annual Sentinels of Safety award to Burra Burra Mine, the first of six wins for TCC mines. All but seven of Burra Burra's 225 workers were present at the banquet to receive their individual certificates. Among those pictured in this photograph are Lonas Payne, Wade Pickelsimer, R.W. Payne, Ben Cantrell, C.A. Suits, and Lon ?.

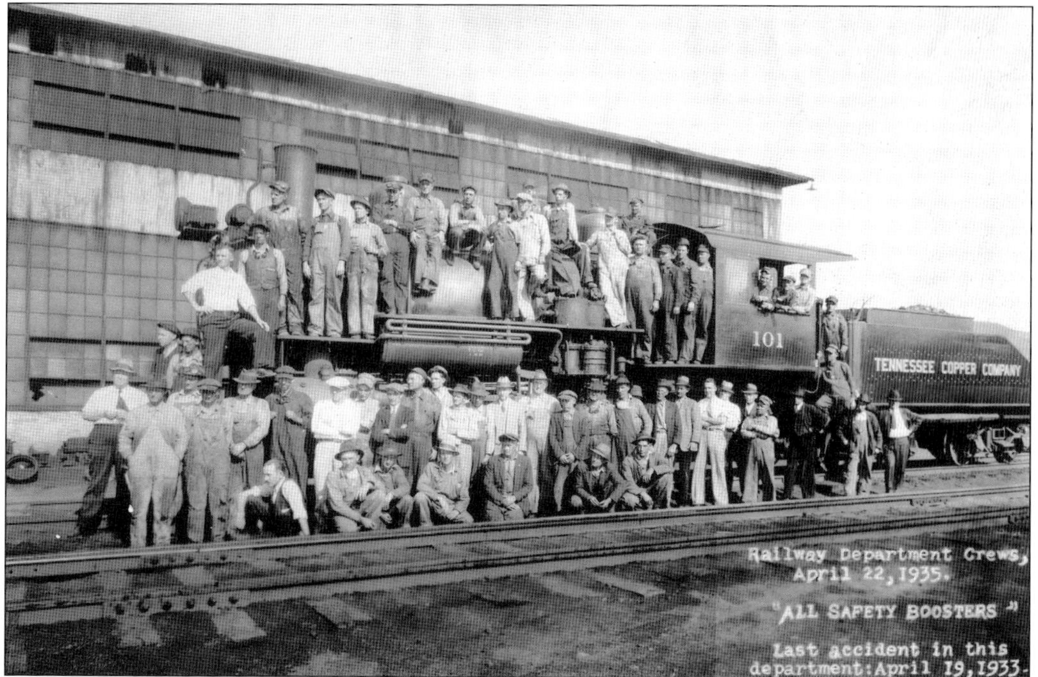

Railroad Department, 1935. It became a tradition in the TCC railroad department to pose for an annual photograph celebrating the department's safety record. The fourth man standing from lower right is conductor W.C. "Chubby" Smith (see page 64), who worked for the railroad department for 43 years.

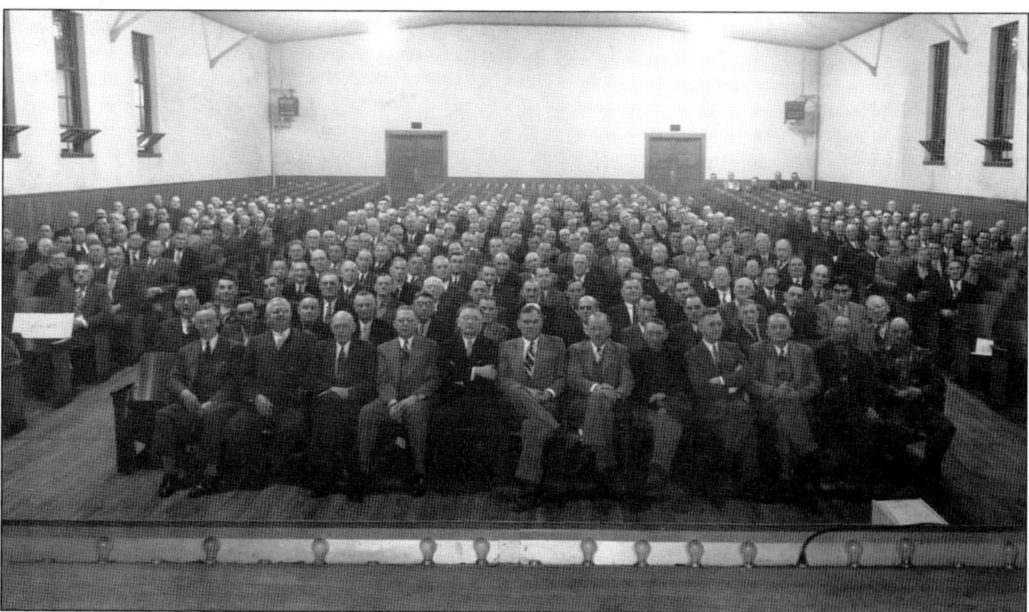

Safety Awards, 1947. On December 22, 1947, several hundred employees gathered for the formal presentation of 20-year safety awards. Second, third, and fourth from left in the first row of this photograph are TCC executives C.H. McNaughton, F.J. Longworth, and T.A. Mitchell. Fourth from right in the same row is safety director S.E. Sharp.

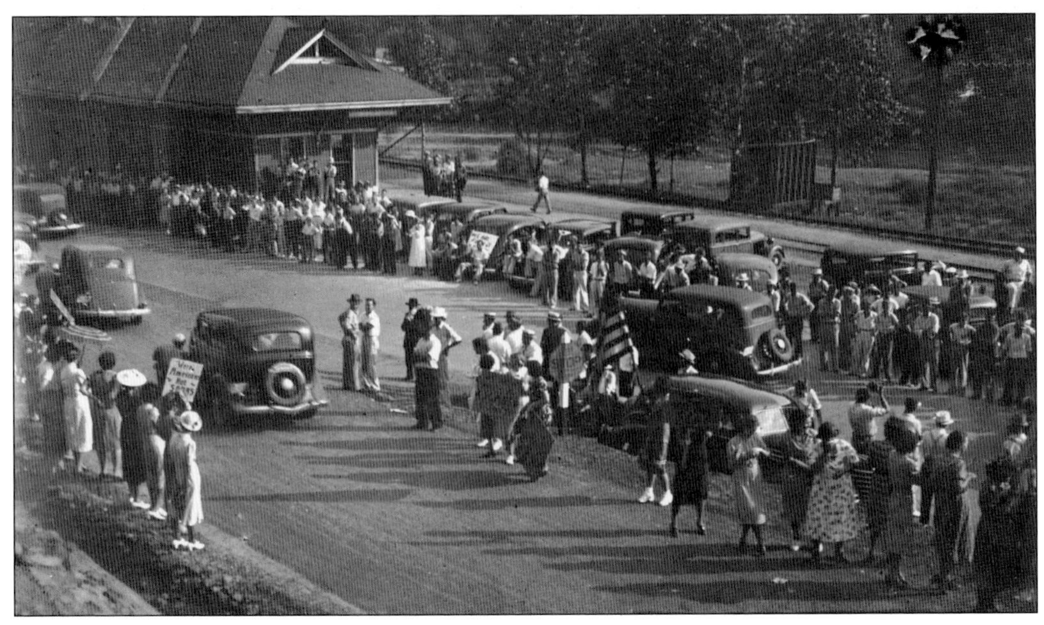

Labor Strike, 1939. In July and August 1939, a six-week strike by 1,400 TCC employees caused the complete suspension of company operations. The strike originated from a struggle for supremacy between the American Federation of Labor, which had represented TCC employees, and the Congress of Industrial Organizations (International Union of Mine, Mill and Smelter Workers), which had represented the former DC&I employees prior to DC&I's purchase by TCC. Although sporadic conflicts broke out again the following year, a compromise was reached in August 1939, and employees returned to work on August 28. Above, strikers and supporters gather at the L&N depot in Copperhill. Below, on the day of the reopening, new men line up outside the Copperhill plant to apply for work.

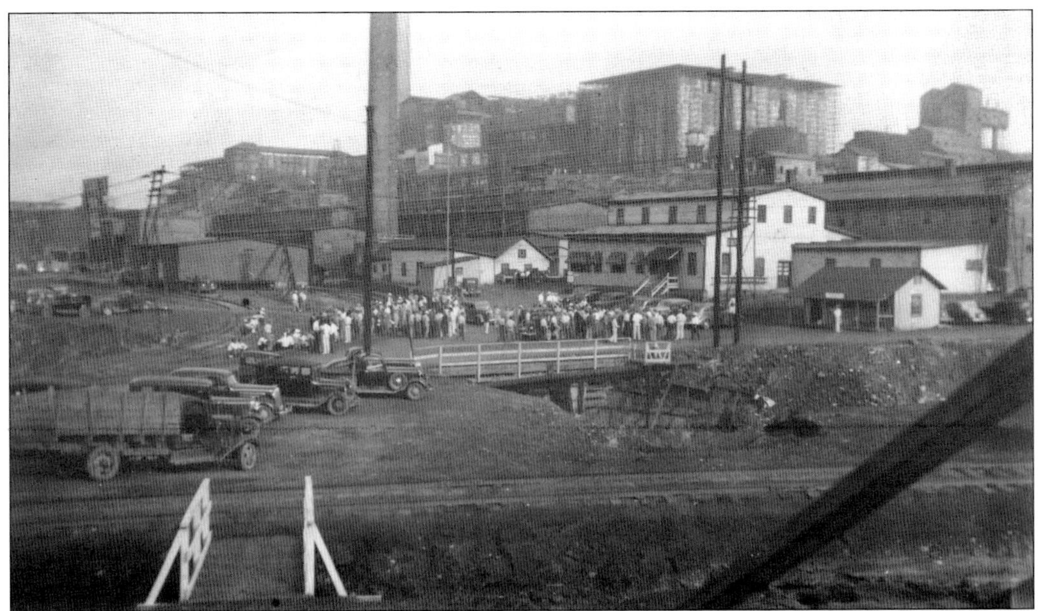

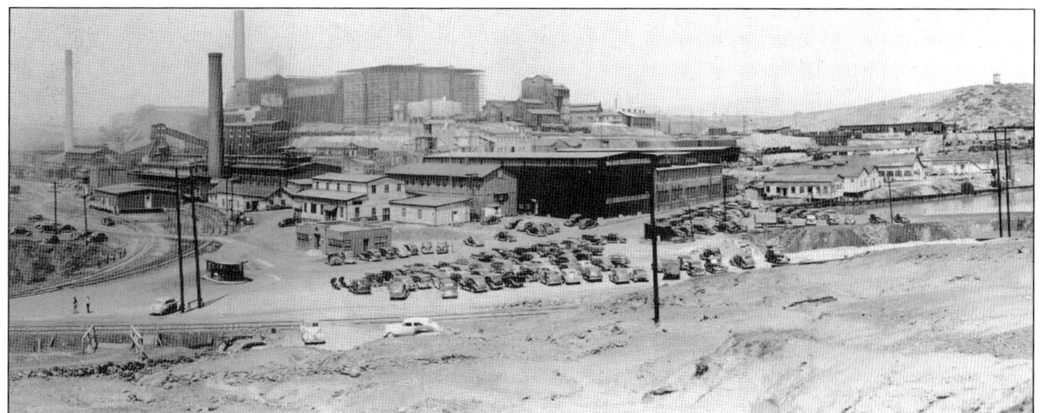

TCC in Wartime, 1940s. In December 1941, TCC began construction of East Tennessee Ordnance Works, a government-owned contact acid plant that would produce 40 percent oleum to make TNT for the war effort. The No. 2 chamber acid plant, visible at left beyond the smokestacks in this c. 1940 photograph, was demolished to make room for the new plant, which was brought online in July 1942 and began shipping to TNT plants all over North America. After the war, with the government contract at an end, the ordnance facility became TCC's No. 2 contact acid plant, which remained in operation until August 1976.

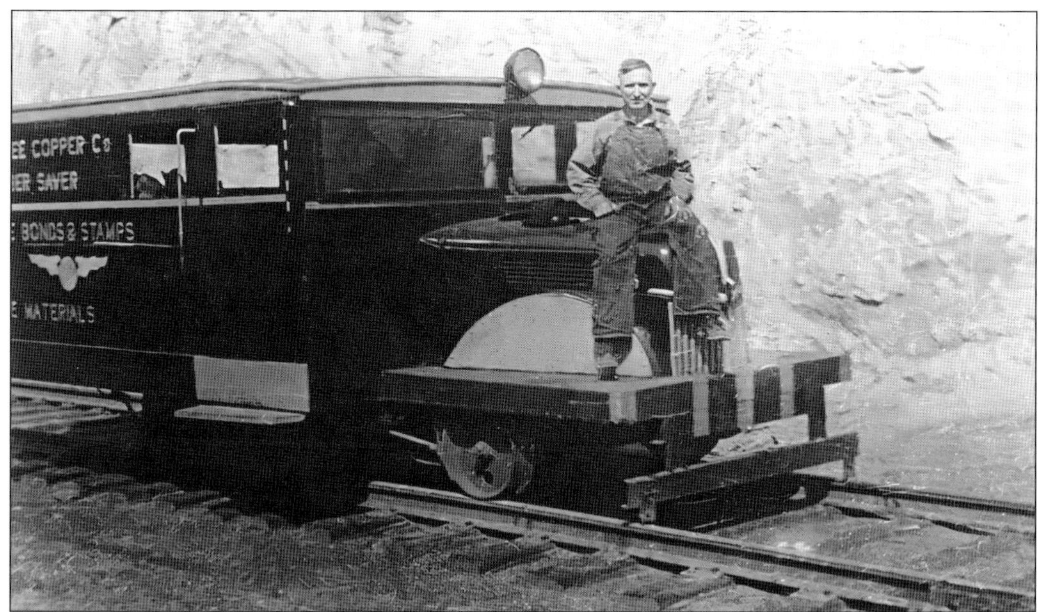

Rubber Saver Train, 1940s. Elisha Panter poses on one of the TCC "Rubber Saver" trains that were put into scheduled service during World War II to help move passengers, mail, samples, and light freight to all parts of the property. There were reportedly two trains, with one based in Copperhill and one at London Mill, both of which ran along the company railroad network. (Courtesy of Thomas E. Panter III.)

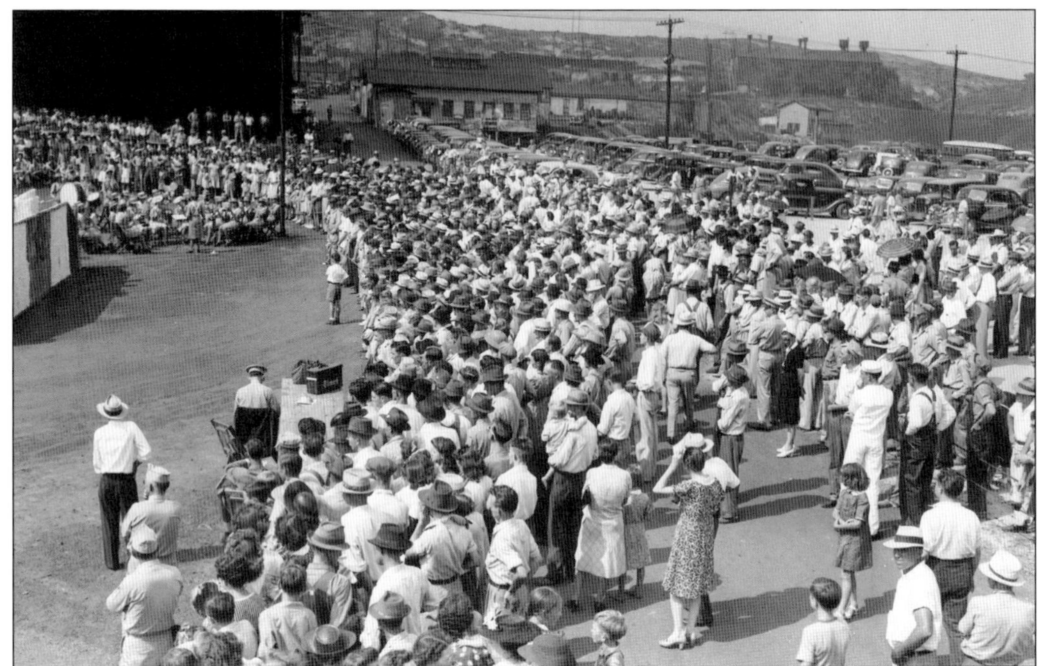

"E" Awards Ceremony, 1943. On August 9, 1943, a crowd of more than 3,000 TCC employees, family members, and other guests gathered near the Copperhill service office for the presentation of twin Army-Navy "E" pennants, awarded for excellence in production, to Tennessee Copper Company and the East Tennessee Ordnance Works. Among those present, in addition to several high-ranking military officers, was future senator and vice-presidential candidate Estes Kefauver, then a US congressman from Chattanooga. Below, before the ceremony, the band of the Third WAC Training Center at Fort Oglethorpe, Georgia, marches along the railroad tracks toward the main gate.

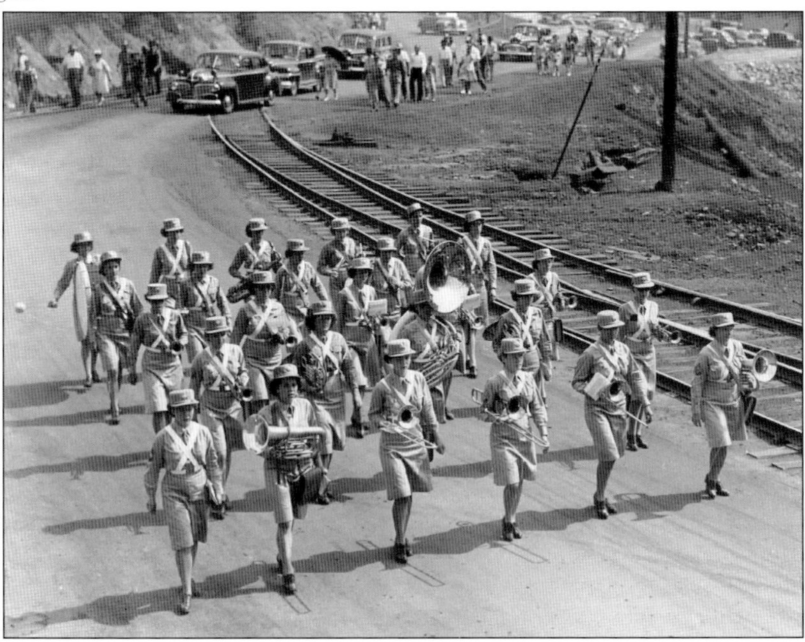

Two

The 1950s

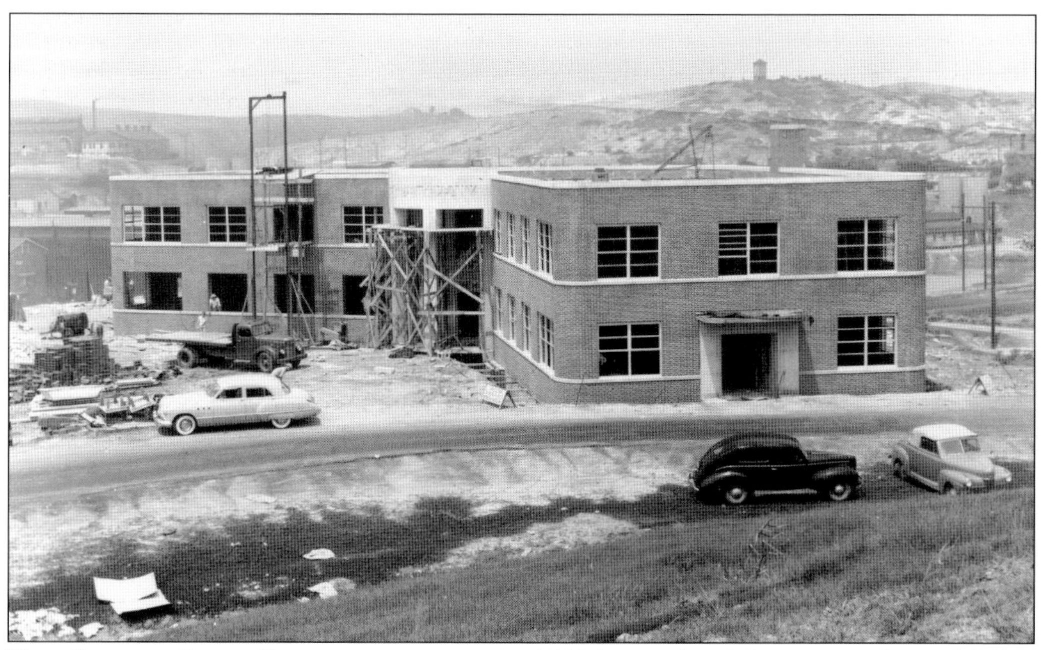

New General Office Construction, c. 1949. TCC began the 1950s in a new General Office building at the site of the original offices in Copperhill. During demolition in the late 1940s, office personnel were moved temporarily to the Blue Goose hotel. The new offices were first occupied in mid-1950 and are still standing. The original Blue Goose was replaced with a modern brick building around the same time.

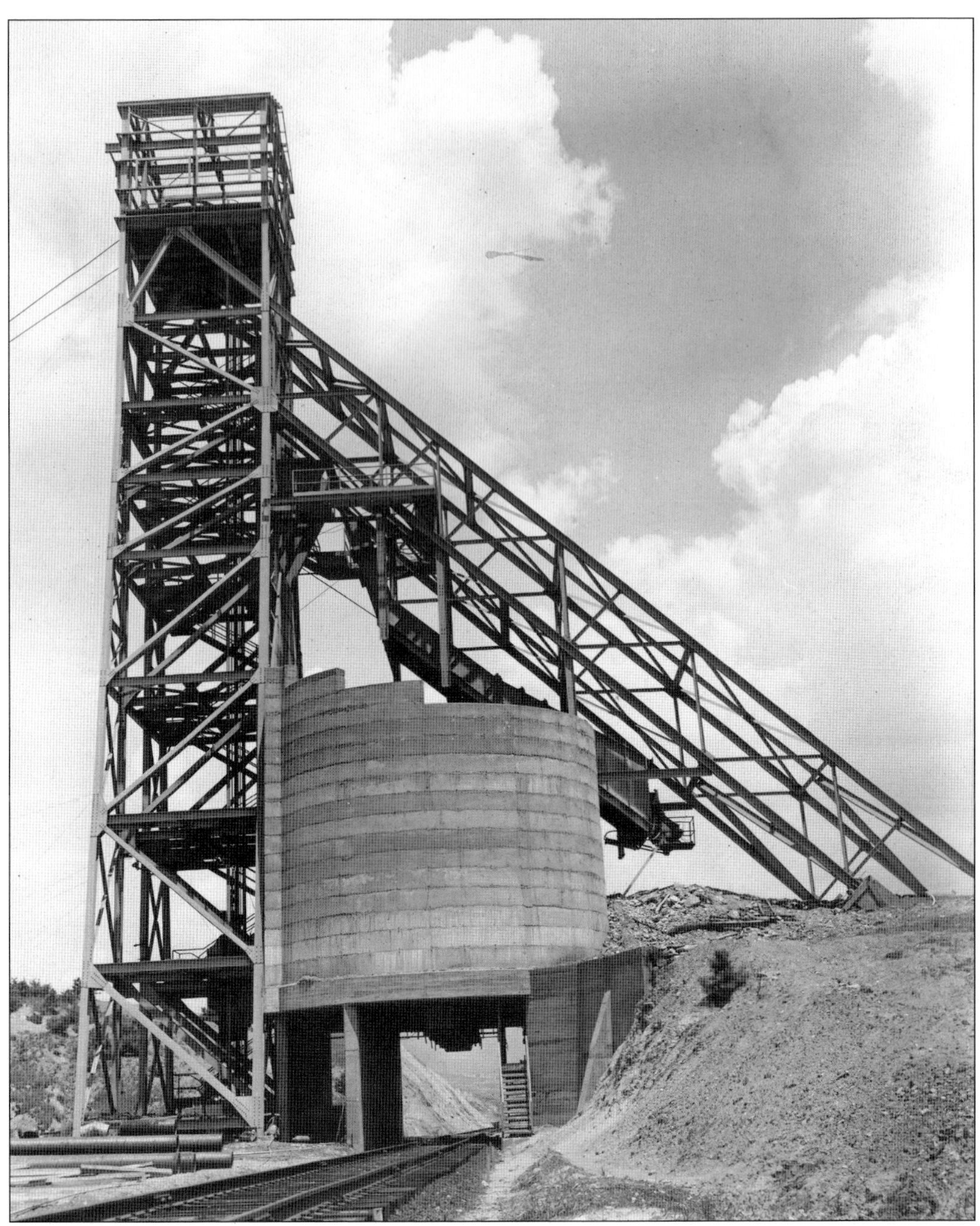

CENTRAL SHAFT, EARLY 1950S. In 1949, TCC began construction of a new shaft and equipment to serve as a centralized hoisting site. During construction, underground drifts were driven to the site from several nearby mines so that ore from multiple mines could be transported to the new shaft for hoisting. The headframe, which was designed by the TCC engineering department and erected by TCC riggers, required more than a half-million pounds of steel. It was put into partial operation in mid-1952 and full operation in May 1953, and began hoisting ore brought by underground rail systems from Boyd, Calloway, and Mary Mines. Its lifting capacity was 450 tons of ore per hour from a depth of 1,400 feet. Its headframe is the only one still standing in the Ducktown district.

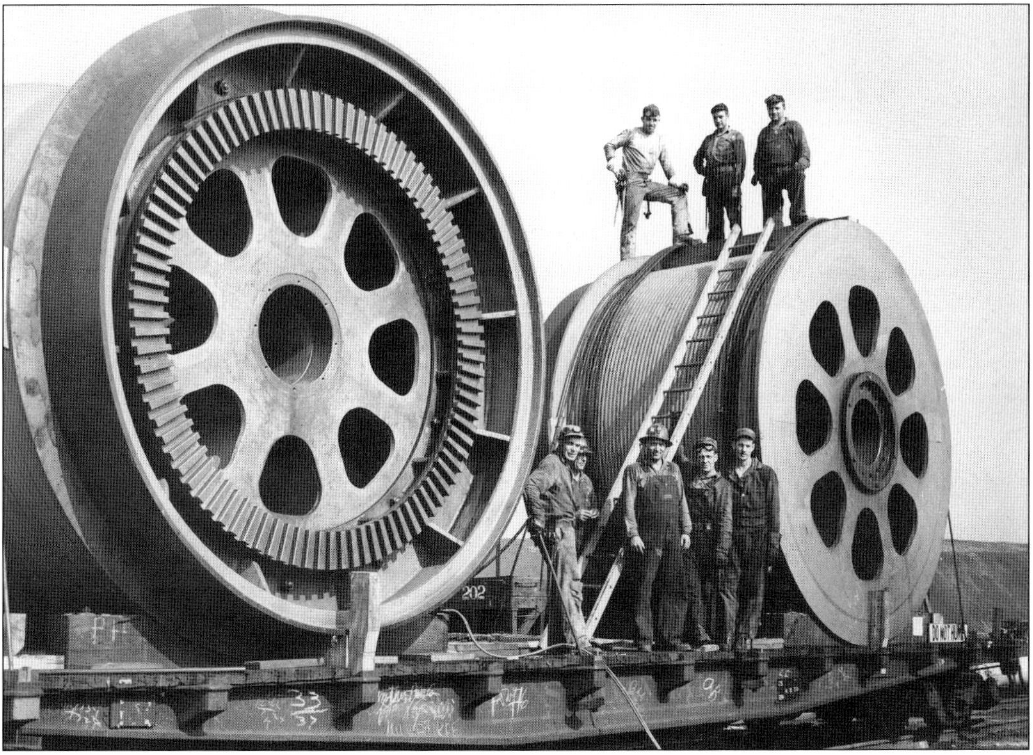

Central Shaft Hoist Drums, 1950. Riggers prepare to unload the hoist drums for Central Shaft. Each drum weighed approximately 43,000 pounds and would use 1.625-inch cables that weighed 10,000 pounds each. From left to right are (on the truck bed) Floyd Chasteen, assistant general foreman Homer Hamby, foreman Worth Franks, Cargus Brooks, and M.A. Collis Jr.; (on the drum) A.T. McCullough, Edwin Smith, and Gordon Mathis.

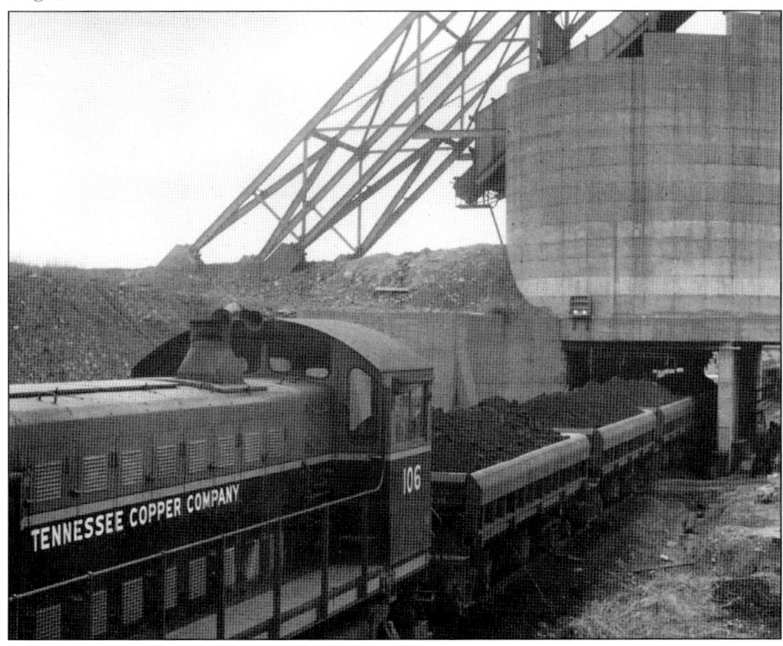

Ore Train at Central Shaft, 1950s. The extensive railroad network that connected all the TCC properties was expanded to the Central Shaft site during the construction phase. The tracks ran under the ore bin, where the ore was loaded into open railcars to be hauled to London Mill for processing.

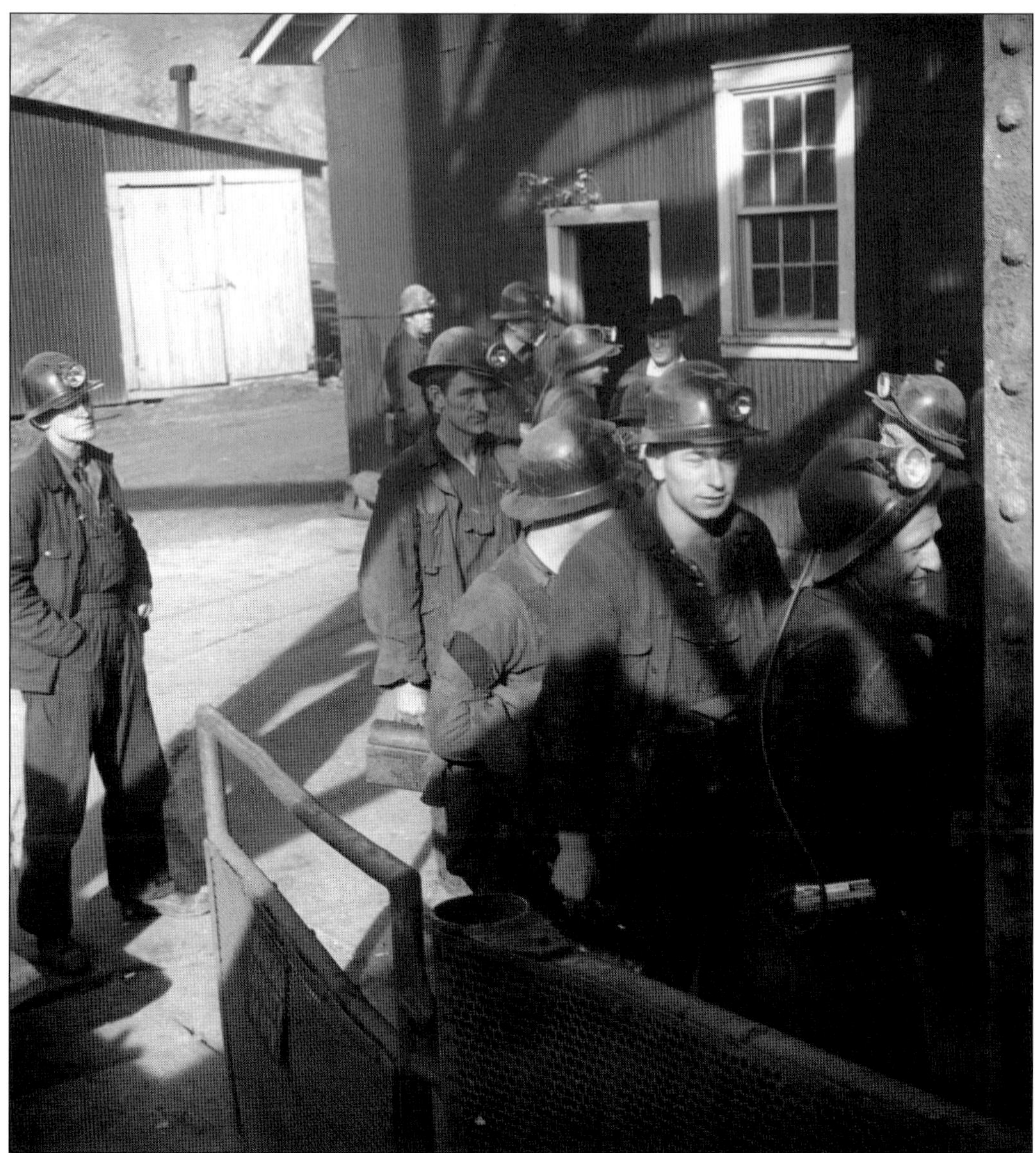

Eureka Mine Shift Change, 1953. In the late-afternoon winter sun, the Eureka Mine headframe casts long shadows over evening-shift miners as they walk toward the man-cage to be lowered down the shaft. At the front of the line at far right is skip conductor Fred Early. In line behind him are drag operators Edward Wilcox and W.D. Huffman, driller Carl Sharp, and stope foreman Pat Hughes. L.M. Newberry, driller and extra foreman, is standing at left. Although it was within sight of the DSC&I/DC&I Isabella plant and mine, Eureka was part of TCC's original property acquisition in 1899. It was worked sporadically for the first quarter-century before being put into small-scale production in 1925. After TCC's purchase of DC&I in 1936, Eureka and Isabella were worked as a single mine, with the ore processed at Isabella Mill.

Safety Board, 1953. Acting shift foreman Frank England writes an instruction to his crew on the evening shift at Eureka Mine. The Eureka safety board was located in the hoist house and was divided into three sections, with one for each shift. Instructions were written in chalk and were erased only after the listed tasks had been completed or the listed conditions had been corrected.

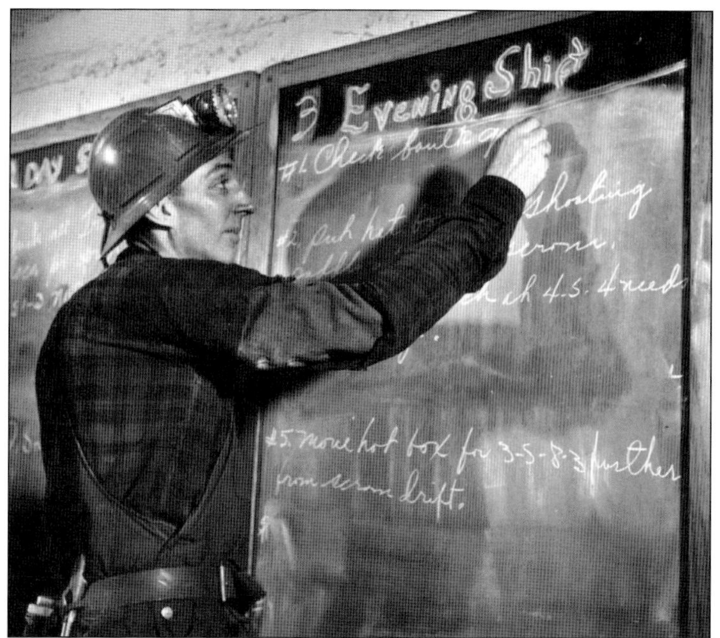

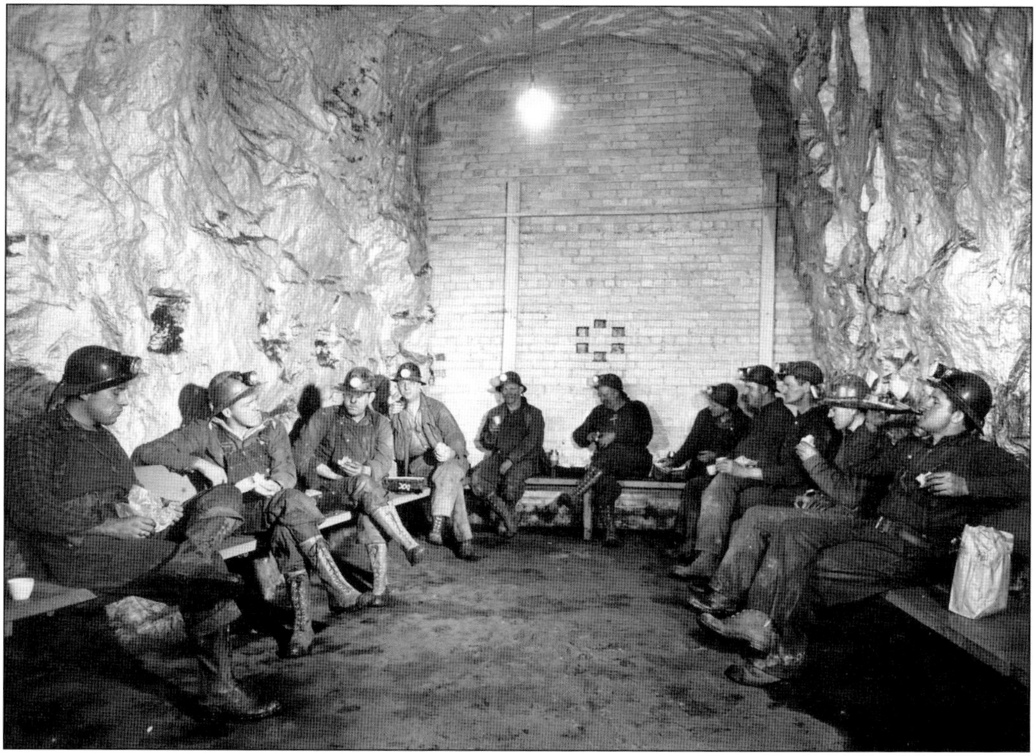

Lunchtime at Eureka, 1953. This third-level room was equipped with a steel door and other protective devices to serve as a safety room in case of emergency. On most days, however, it served primarily as a comfortable place to eat lunch. From left to right are E.E. Hensley, Roy Graves, Grover Buchanan, W.L. Hensley, G.M. Ledford, E.A. Newberry, W.T. Fowler, E.F. Bryant, J.C. Bryant, James W. Holder, and G.M. Payne.

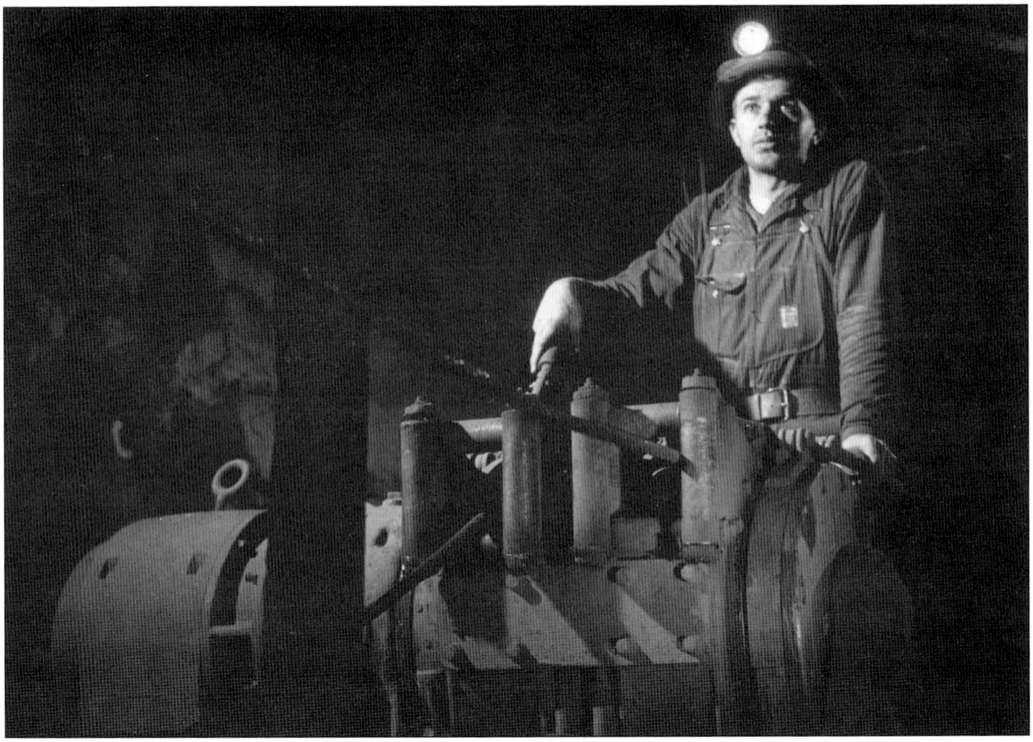

EUREKA MINE WORKERS, 1953. Above, dragman Charlie Waters operates a drag motor, which required him to keep both hands busy at the controls. The drag, pulled by an electric motor capable of moving large quantities of ore, was a considerable improvement over the older method of moving and loading ore by hand. Below, W.B. Bandy operates the skip control board, which functioned as a counting device to keep track of the number of skips of ore hoisted to the surface.

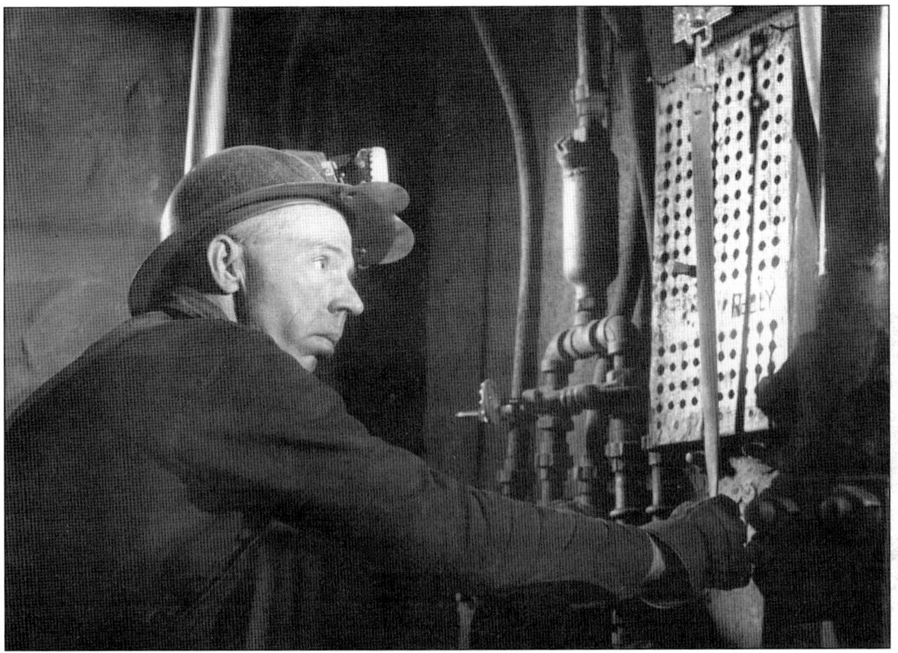

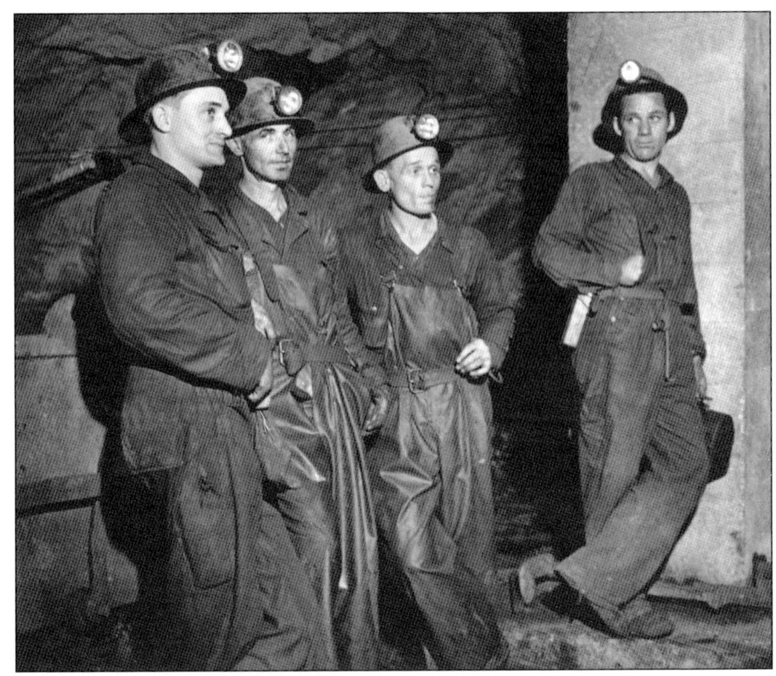

Calloway Drillers, 1954. From left to right, drillers Robert Talley, Johnny Ray, Kenneth Stover, and Bill Dean wait for the man-cage on the 10th level of Calloway Mine. They would be lowered down the mine's new "B" shaft to continue the work of development drilling at a depth of about 2,000 feet. The new shaft ran from the 10th to the 24th level underground but was not open all the way to the surface.

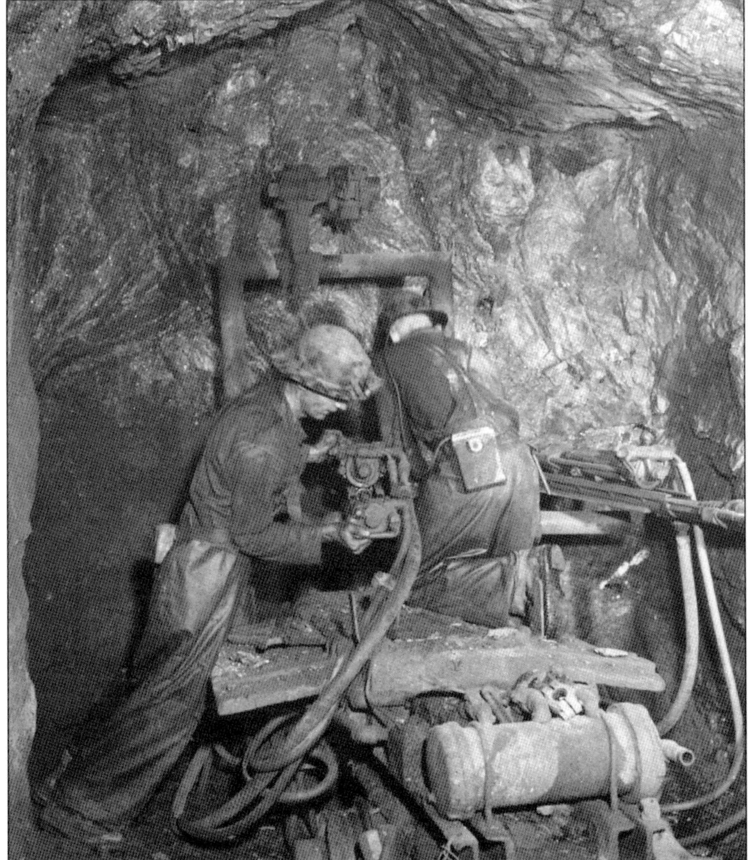

Drilling Jumbo, 1954. By the 1950s, mining was no longer a matter of picks and shovels. In August 1954, Knoxville photographer Emil Willimetz captured this image of two miners assembling and setting up a drilling jumbo to be used for rock drilling in Calloway Mine. Jumbo drilling was first developed in the 1930s during the building of Hoover Dam but soon came into common use in the mining industry.

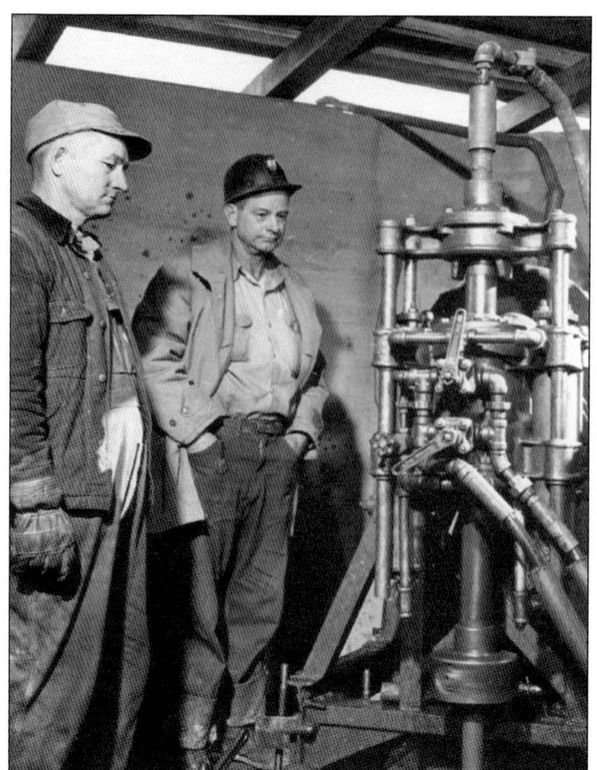

Extending the "B" Shaft, 1956. When exploration revealed that the newly opened section of the Calloway ore body was far larger than TCC had realized, engineers decided that the underground "B" shaft needed to be extended all the way to the surface. In this photograph, diamond driller Cecil Bruce (left) and shaft sinking foreman E.H. Hawkins prepare for the drilling of a pilot hole to help facilitate the new construction.

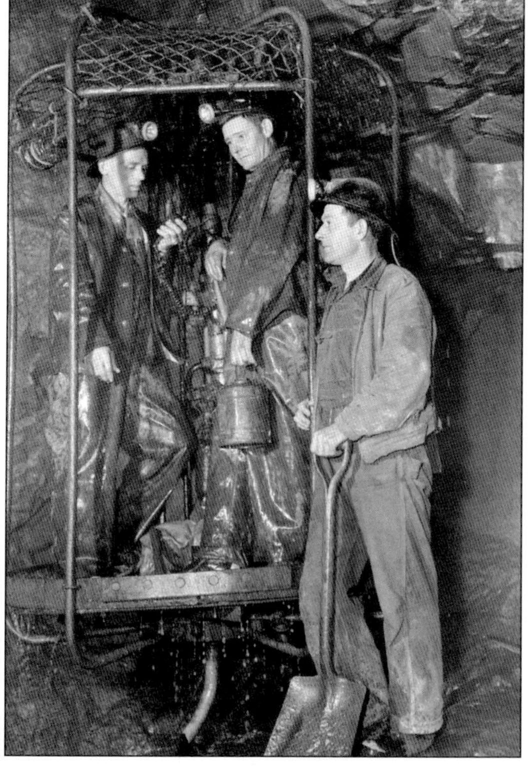

Calloway Drillers, 1956. Calloway "B" shaft drillers Kenneth Stover (left) and Harvey Bell stand in the "birdcage," which was raised and lowered through the four-inch pilot hole that had been drilled down from the surface. Their job was to expand the pilot hole to a shaft 6 feet in diameter that could then be drilled and blasted to a final diameter of 12.5 feet. At right is Bill Cearley.

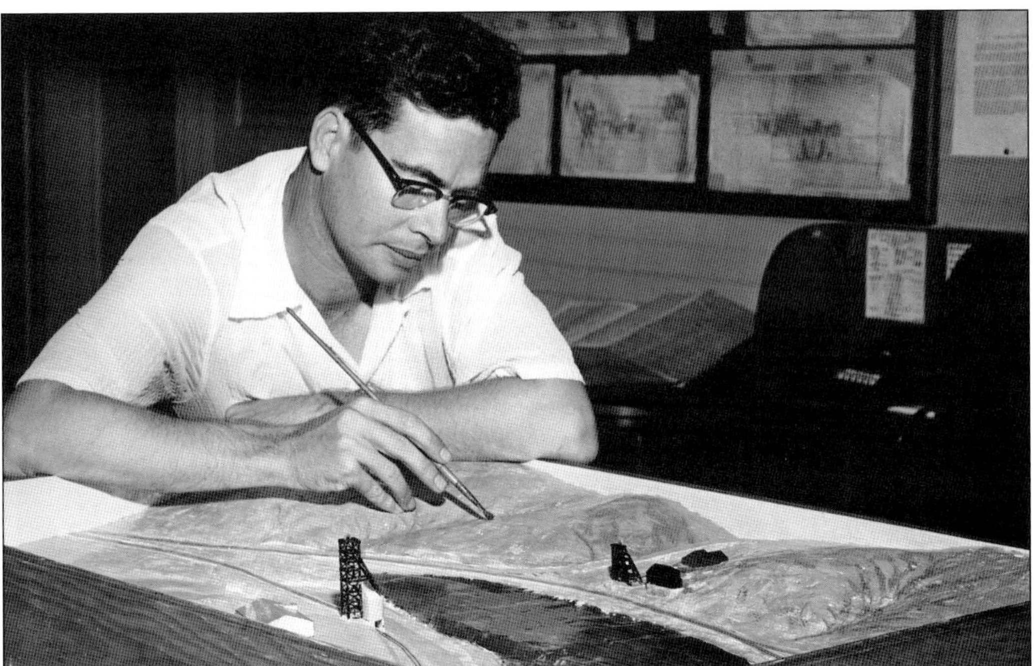

"B" Shaft Construction. Above, in 1956, structural engineer David Bell puts the finishing touches on a model of the Calloway site that would be used as a visual aid in planning details of the surface structures and roads that would be required. The model, built to a scale of fifty feet to one inch, was so detailed that it was often consulted by TCC's mining engineers in order to save sketching time or trips to the actual site. Below, at the Calloway site in May 1957, riggers use a motorized crane to unload half of a hoist drum, weighing 26 tons, that would be fastened to the other half using a 15-ton shaft. It was the largest and heaviest hoist drum in use at any TCC mine to date.

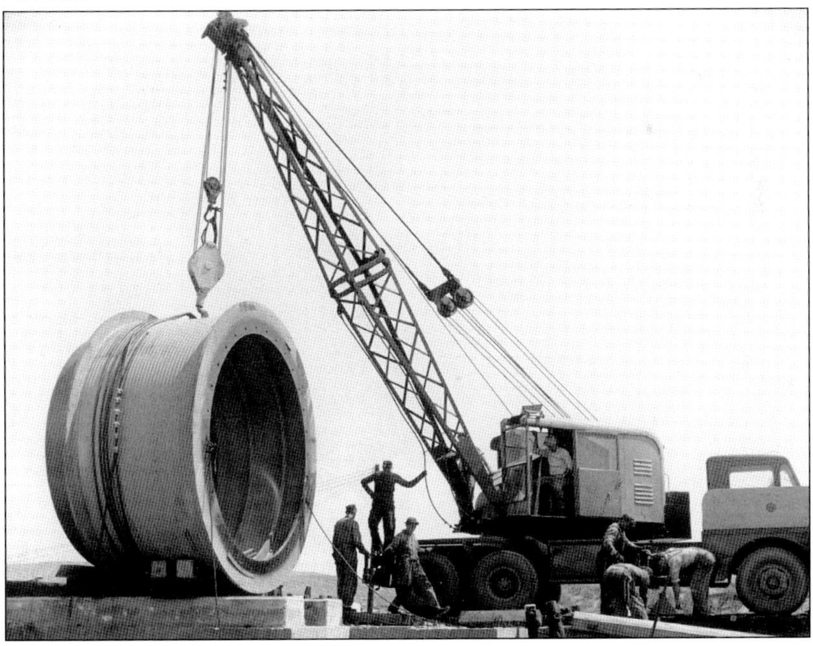

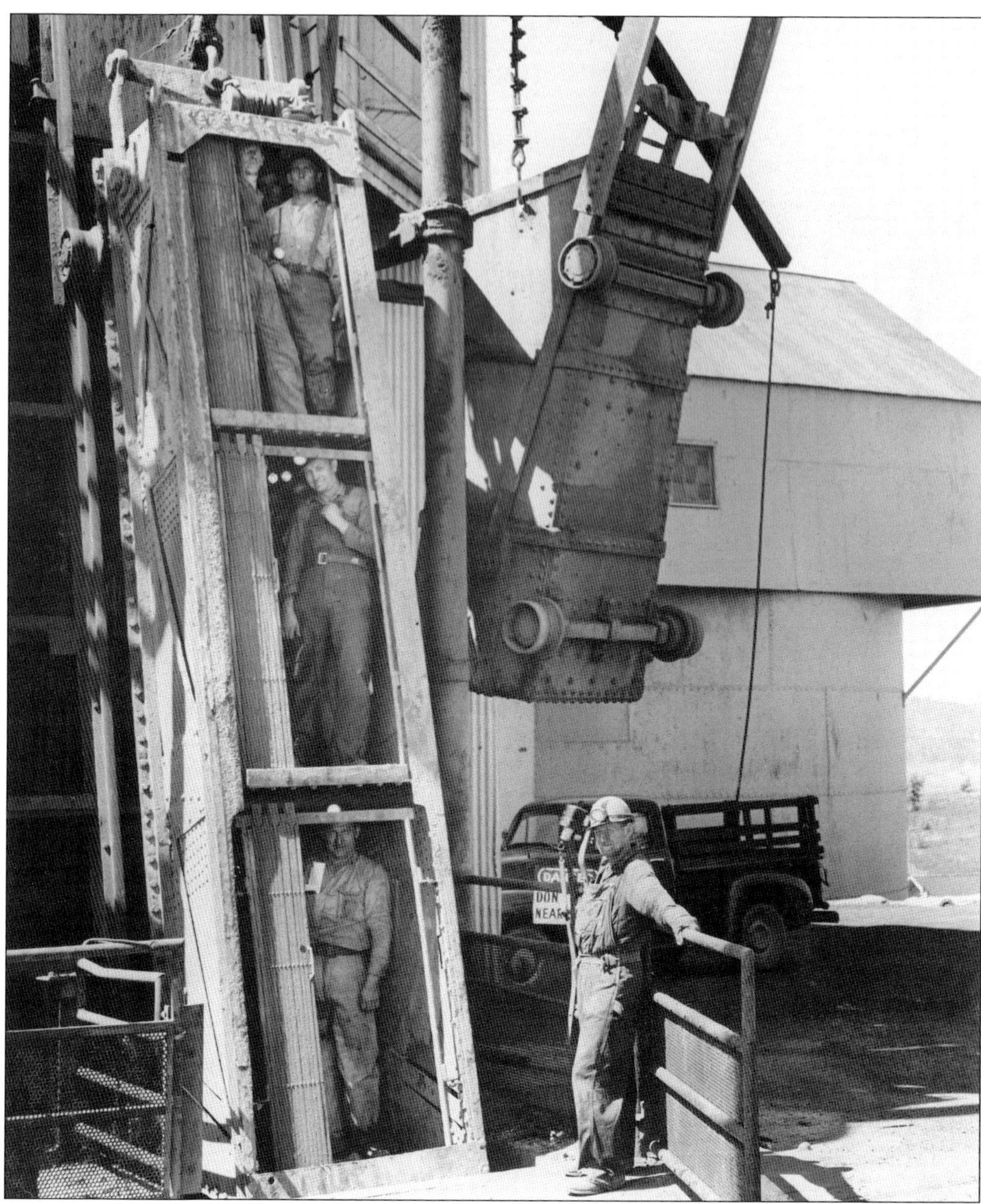

MAN SKIP AT BURRA BURRA, 1958. On May 28, 1958, after almost six decades of continuous operation, the last drilling and blasting was done at Burra Burra Mine. The hoisting of ore continued for two more weeks, and the salvaging of equipment such as drags, fans, pumps, motors, and tram cars took about another month. Then the pumps were stopped, and the mine was allowed to fill slowly with water. This photograph, published in June 1958, was taken on one of the last trips of the man skip. The ore skip hangs at upper right. The rich lode at Burra Burra had been mined to a depth of about 1,600 feet, and some ore still remained at a deeper level, but TCC had determined that attempts to extract it would not be technologically or economically feasible.

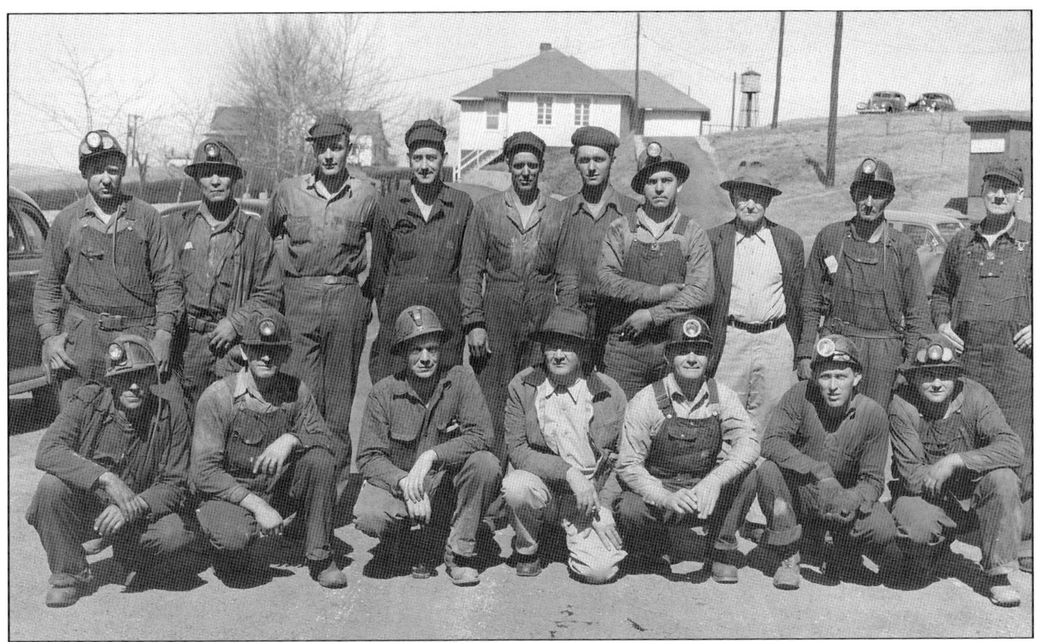

Miners at Burra Burra, Early 1950s. Third from left in the first row of this photograph, set against the background of the TCC mine office, is mining department superintendent H.F. Kendall. Fourth, fifth, and sixth from left are ? Amburn, Clifford Helton, and J.R. Ramsey. In the second row, Wayne Mashburn is third from left, Kenneth Hughes is sixth from left, and Sam Craig is eighth from left.

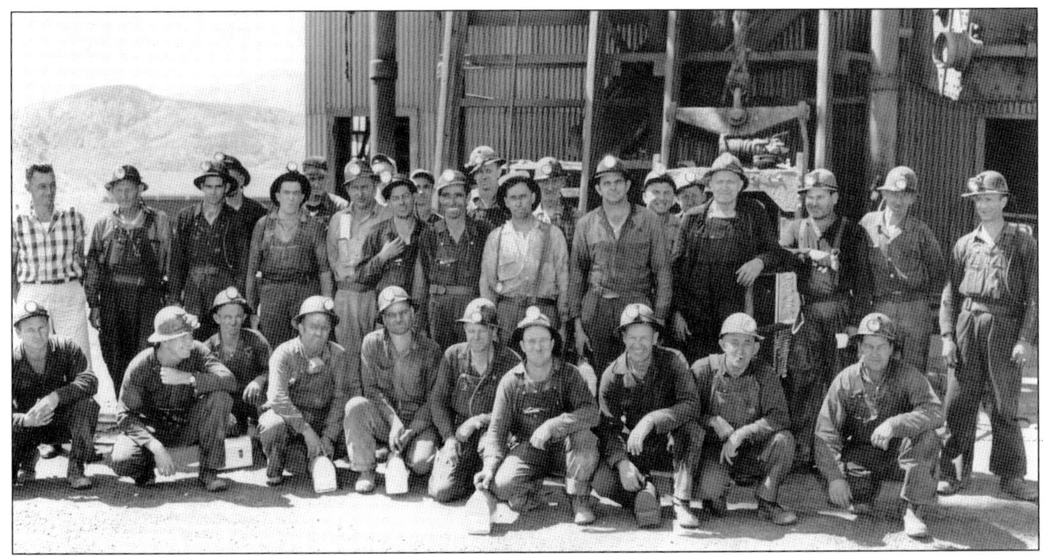

Last Shifts at Burra Burra, 1958. On May 28, 1958, the day and evening shifts at Burra Burra posed for this final crew photograph. The last ore was actually hoisted from the mine two weeks later, on June 13. It was the end of 59 years of continuous operation for the historic Burra Burra Mine.

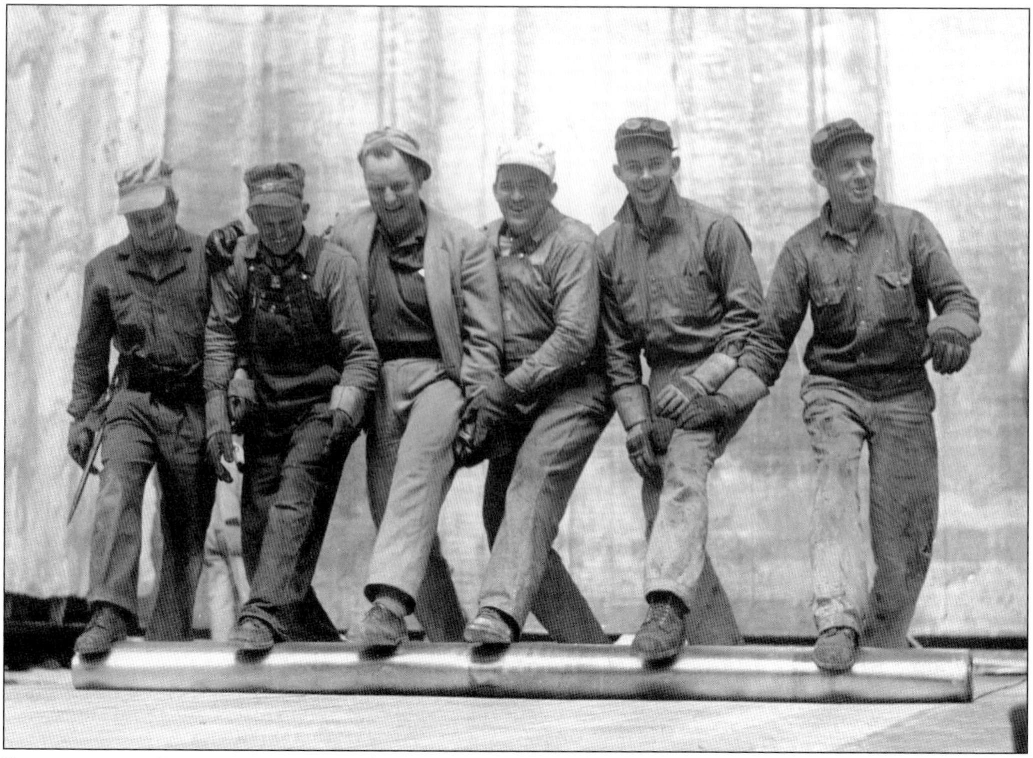

Rolling out Lead, 1953. From left to right, riggers Gordon Mathis, Solon Simonds, Harry McNally, Roy Reiordan, John Turner, and Winston Hawkins demonstrate how sheets of lead were rolled out for use in rebuilding a curtain wall in the lead chamber acid plant at Copperhill. The rolls were 36 feet long and 8 feet, 6 inches wide, and weighed 1,836 pounds.

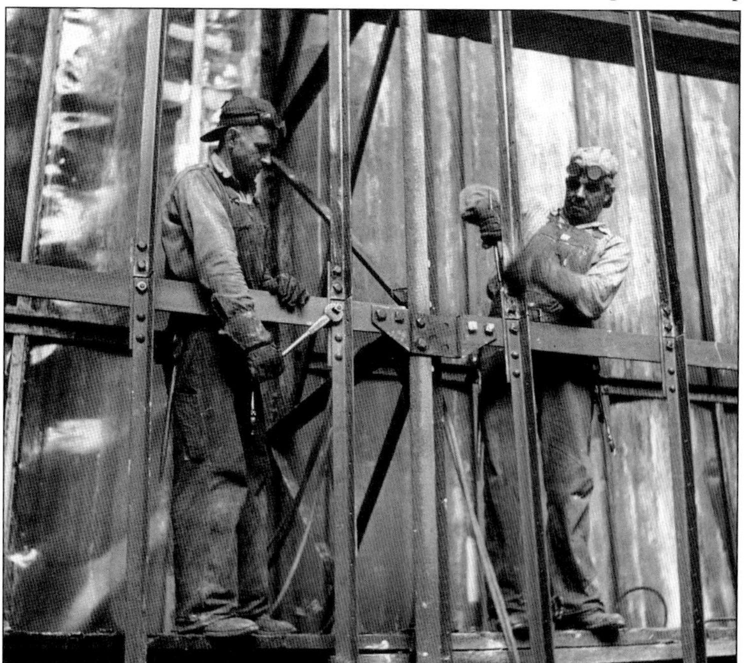

Riggers, 1954. While the lead-rolling riggers were choreographing their chorus line and preparing for their big moment on camera, considerable work was going on elsewhere during the rebuilding of the acid chamber plant. In this early 1954 photograph, riggers Willard Morrow (left) and Prince Johnson use spud wrenches to tighten the bolts that would hold the new vertical steel girts together.

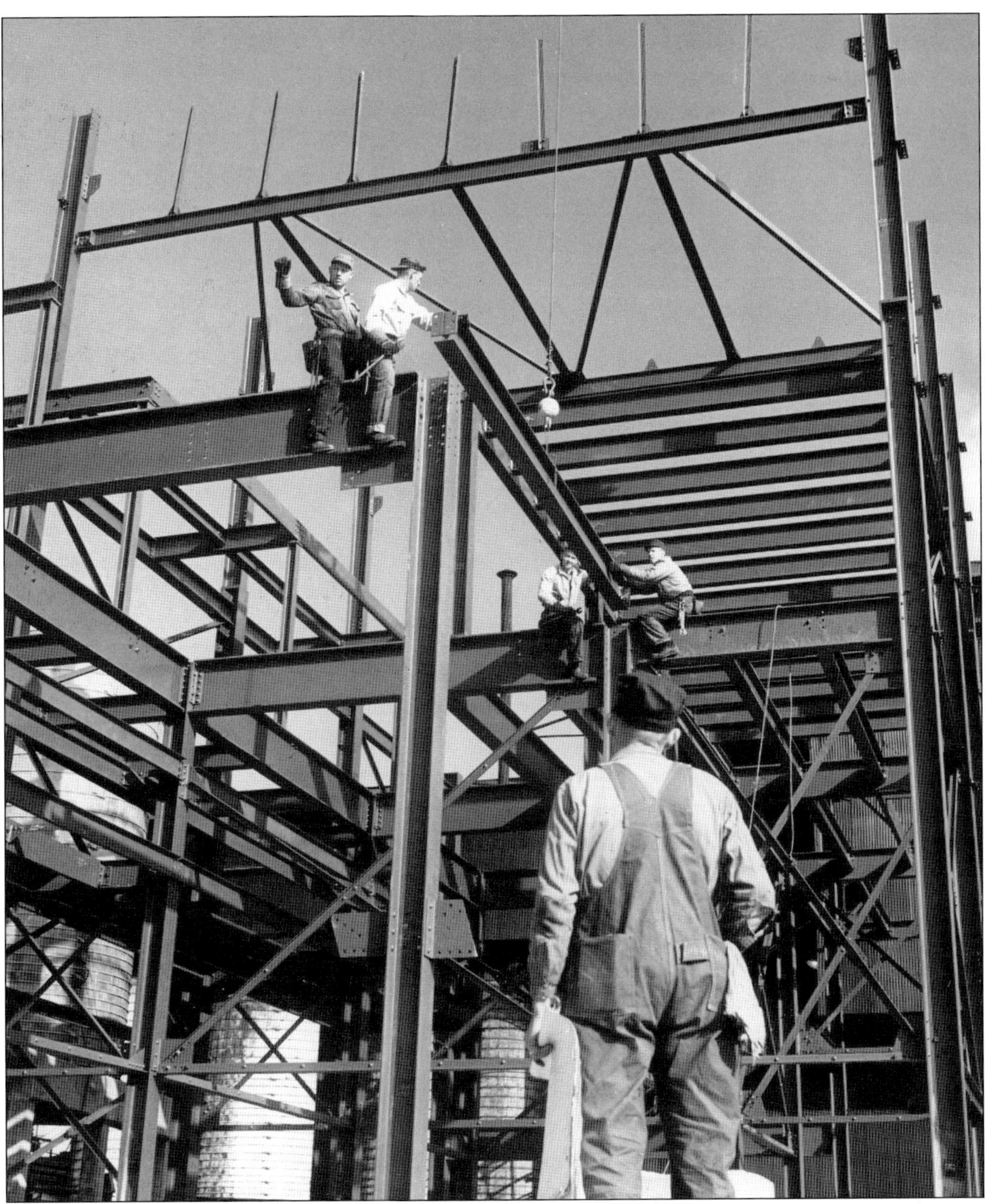

Sinter Plant Construction, 1954. At the same time the acid chamber plant was being rebuilt, work was also beginning on a new sinter plant at Copperhill. In this photograph, rigger foreman Solon Simonds, standing in the foreground, oversees the construction of the steelwork for the new plant as members of a rigger crew direct the landing of an I-beam. At top left, Paul Green signals the operator of a nearby motor crane, and Winston Hedden, beside him, helps guide the beam into place. Helping ease the new beam into its correct position on the other end are John Turner (left) and Winston Hawkins. The steel would be temporarily bolted together by the riggers before being assembled permanently by riveting. The TCC smelting department had done the basic design for the new plant, which increased the company's output of sinter to be supplied to the iron and steel industry.

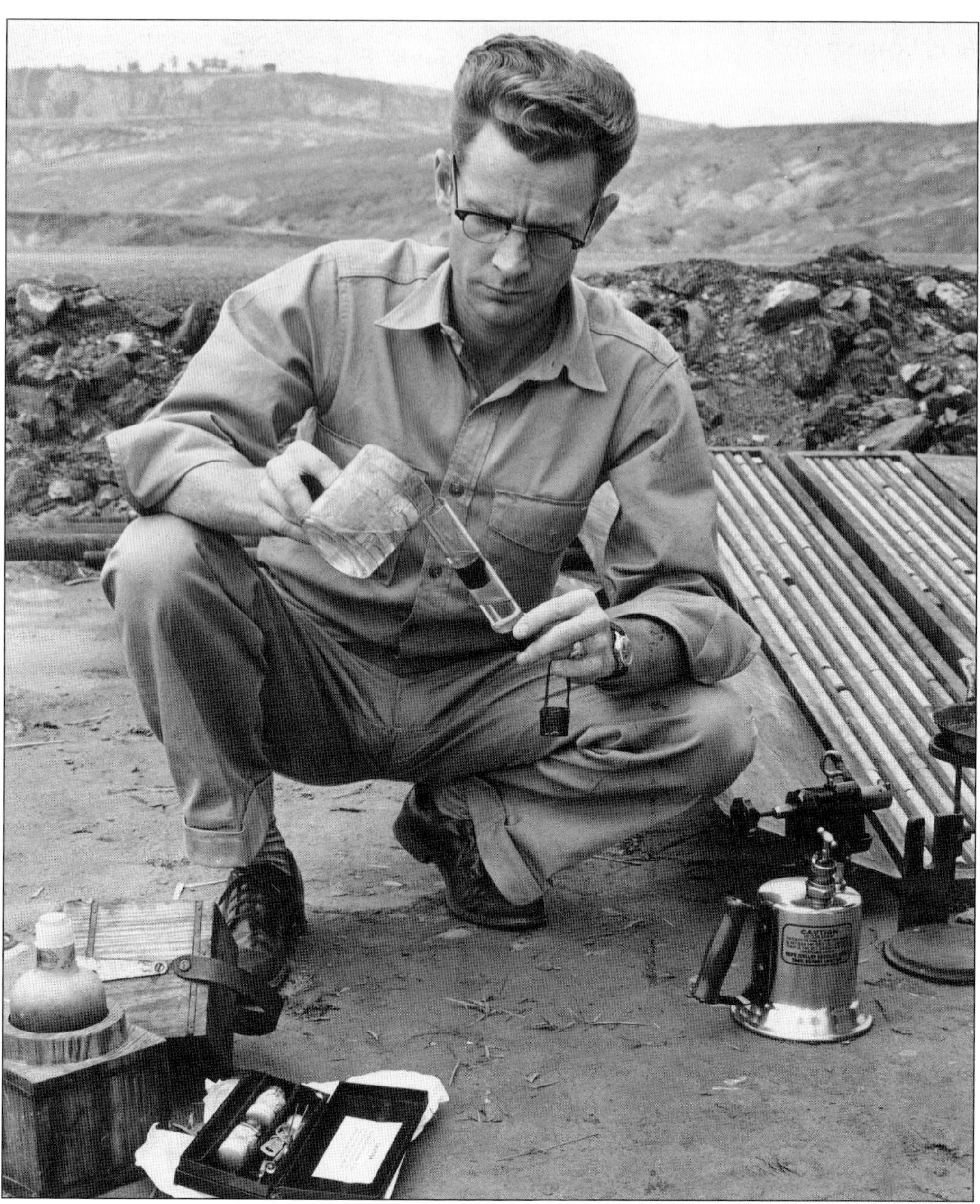

Drilling Location Survey, 1955. For every miner working underground at TCC in the mid-20th century, there were roughly seven employees working on the surface. In this photograph, Hubert Jenkins of the geology department prepares a glass tube to be used in a survey of a new drilling location. The two-section tube was filled partly with warm gelatin, into which a compass needle apparatus was inserted, and partly with dilute hydrofluoric acid. The tube was then lowered into the drilling hole and allowed to stay there for a period of time. When it was retrieved, the compass would be fixed into position by the hardened gelatin, and the inclination of the hole would be marked on the other part of the tube by the etching effects of the acid. The information obtained during the survey regarding the direction and dip of the hole would then be plotted on the geologists' maps.

Acid Loading, 1957. In 1957, TCC made a major upgrade to its acid-loading facilities at the Copperhill plant. The new facilities consisted of three tracks, each with nine loading points, and allowed for the loading of different strengths of acid at the same time. In this photograph, loader Gene Graham is at his station on the left, while inspector Bob Bruce, at right, prepares to check a car.

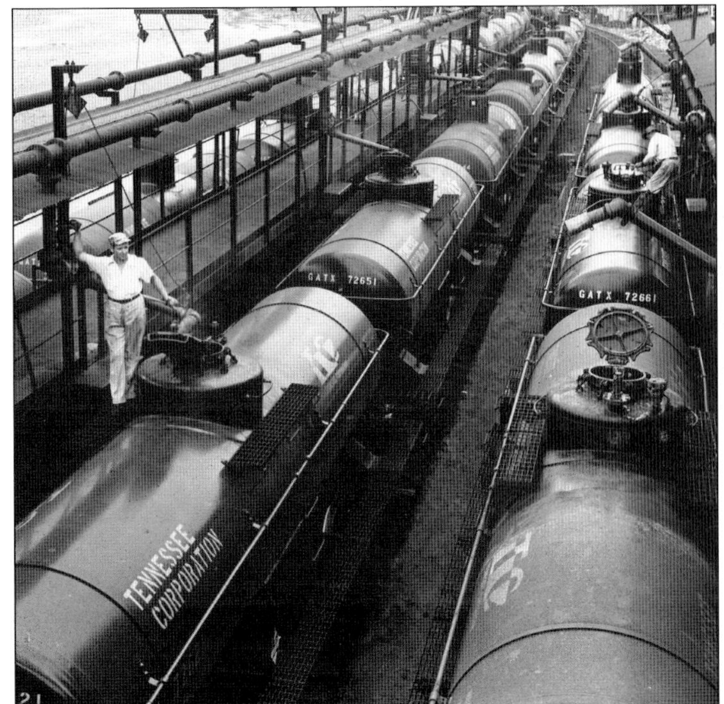

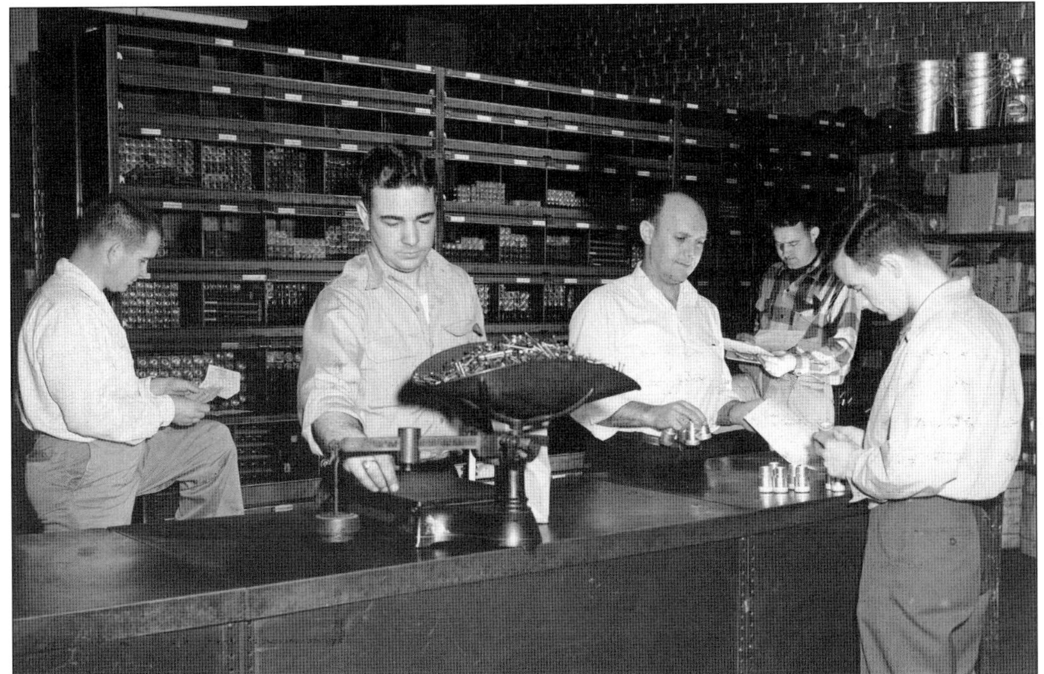

Burra Burra Supply Department, 1955. Whether it was a handful of bolts or a major piece of equipment, the job of the supply department was to make sure every employee and department had the materials to get the job done. In this photograph, Nelson Severinghaus Jr., at right, inspects diamond drill bits. Behind the counter, from left to right, are W.E. Smith, Kenneth Ross, supervisor Earl Bell, and Bill Dunn.

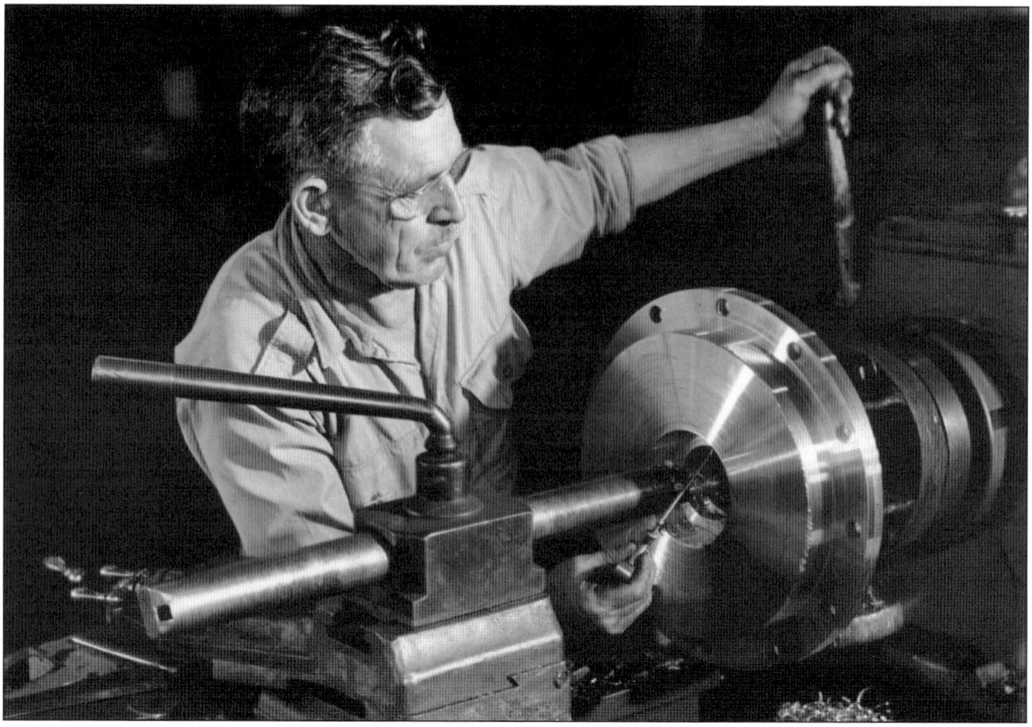

Machine Shop, Early 1950s. Precision was key in the TCC machine shops, where a great deal of the company's equipment was fabricated and repaired. Above, Copperhill machinist Oliver Tilson concentrates on the operation of his lathe. Below is machinist Earl Craig at a workbench in the Copperhill shop in 1952.

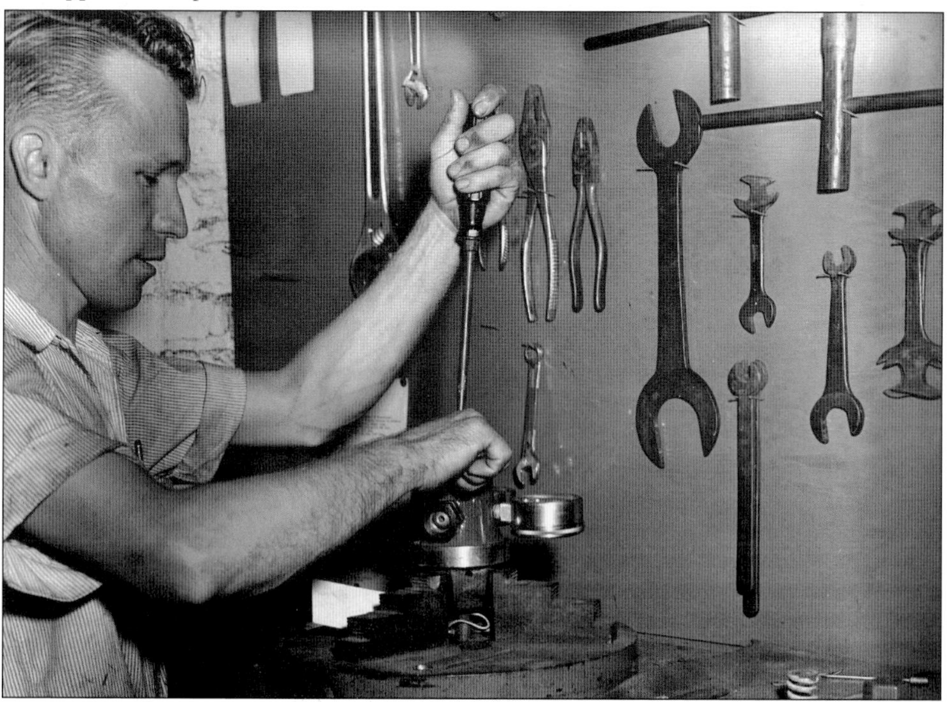

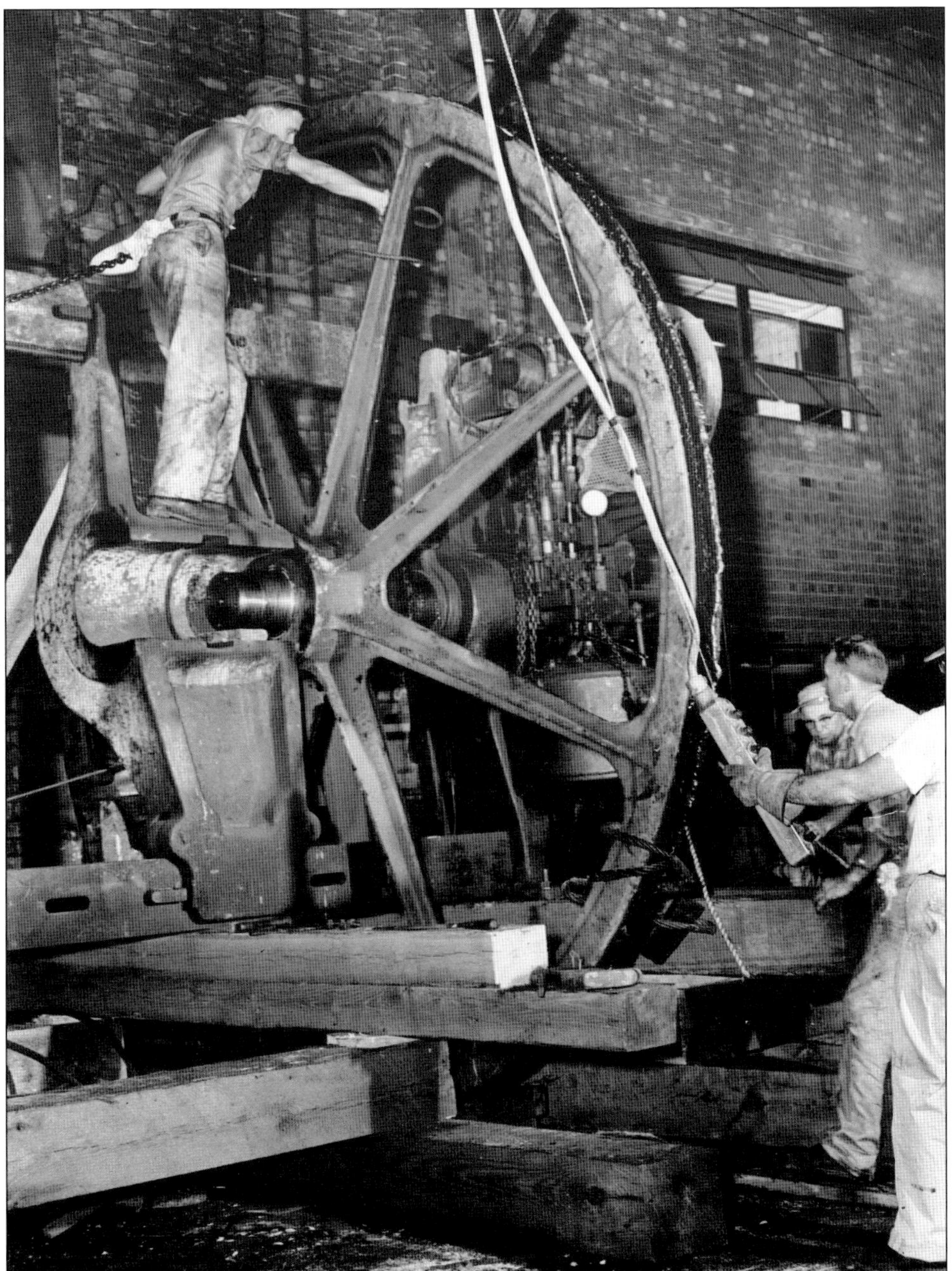

Sheave Wheel Repairs, 1958. In June 1958, the sheave wheel at Central Shaft suffered its own on-the-job injury. The wheel, which was 13 feet in diameter and stood more than 100 feet above the ground, was lowered by motorized crane and brought to TCC's Copperhill machine shop for replacement of a damaged axle. Since the shop's hydraulic-ram wheel press had not been designed for wheels of this size, both the press and the wheel needed to be jacked up and supported with timbers, allowing the wheel to be repaired on-site rather than sent to a larger machine shop in Chattanooga. In a combined operation involving not only machinists but also riggers, truck drivers, and mining department employees, the job was completed in a single weekend with minimal interruption to the hoisting of ore at Central Shaft.

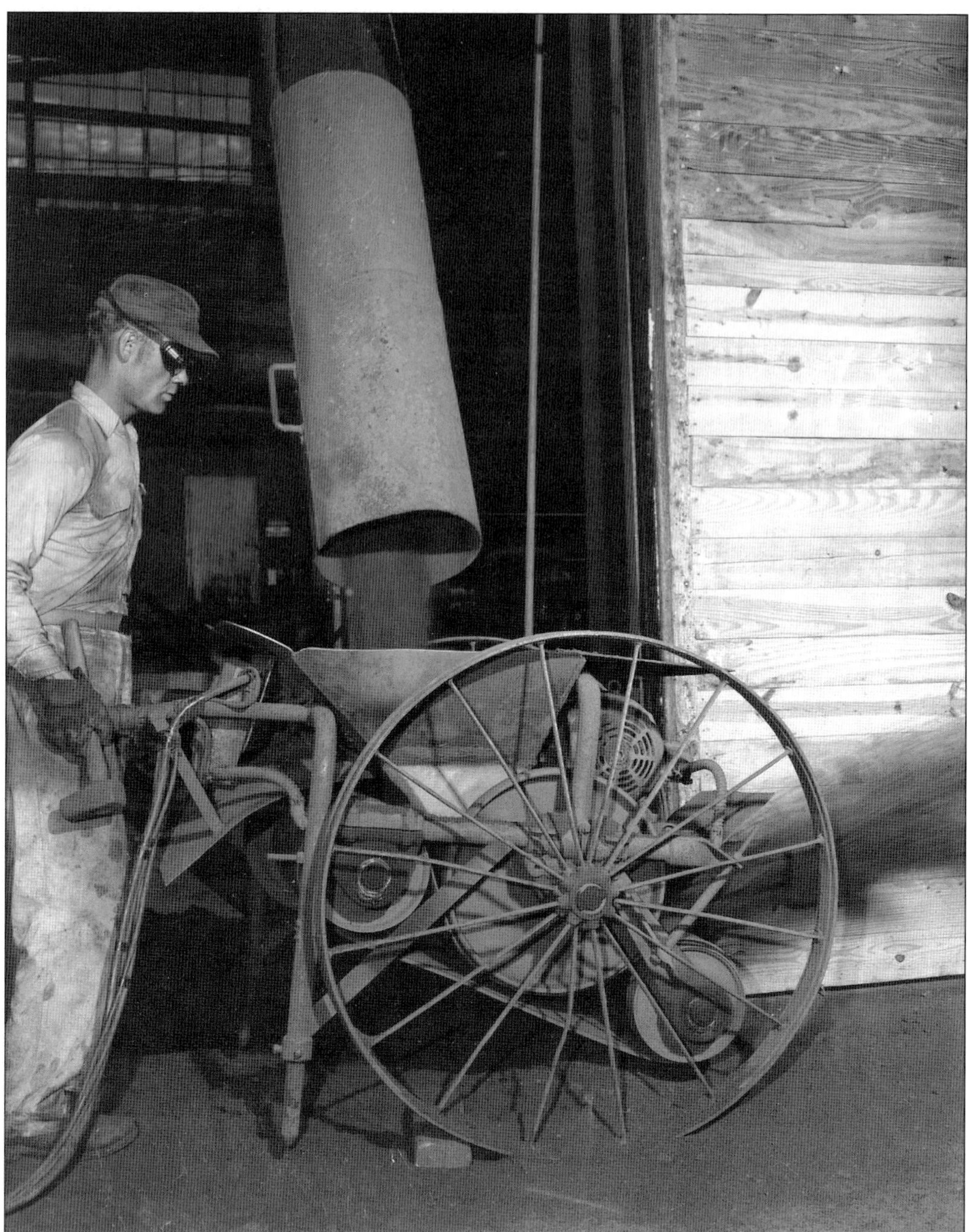

LOADING ZINC CONCENTRATE, 1957. Among the several components of the local ore were iron, which could be sold as iron sinter, and zinc, which had to be removed from the mix in order to make the iron sinter marketable. The process for isolating the zinc during the flotation process was developed at London Mill in 1927 and improved over the years, allowing the company to tap into the market for zinc concentrate. Refined zinc was in demand for galvanizing, as well as for the making of brass, and was beginning to prove useful in die-casting. In this photograph, John Dockery operates the loading machine that transferred the fine granules of zinc concentrate into boxcars for shipment.

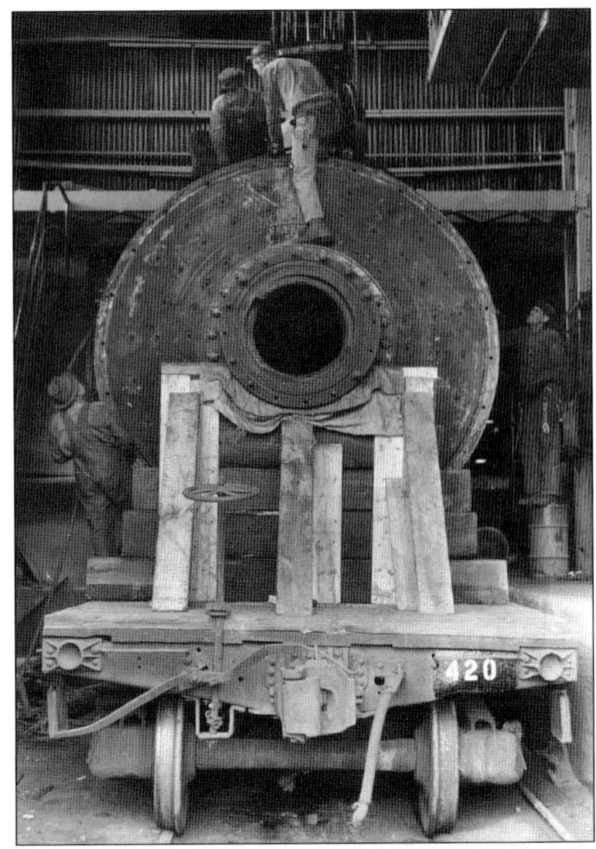

London Mill, 1950s. At right, workers prepare to unload a rod mill at the London Mill flotation plant. The cylindrical structure was filled with long, slender rods that were used for the crushing of ore. The worker at top left has been tentatively identified as Willard Davenport. Below, T.C. Carroll operates the zinc flotation machine. By the time this photograph was taken in mid-1957, TCC was able to extract and ship 63 percent of all the zinc in ore processed at London Mill. No zinc was recovered at Isabella Mill, because the Eureka Mine ore treated there was too low in zinc content to make its extraction commercially viable.

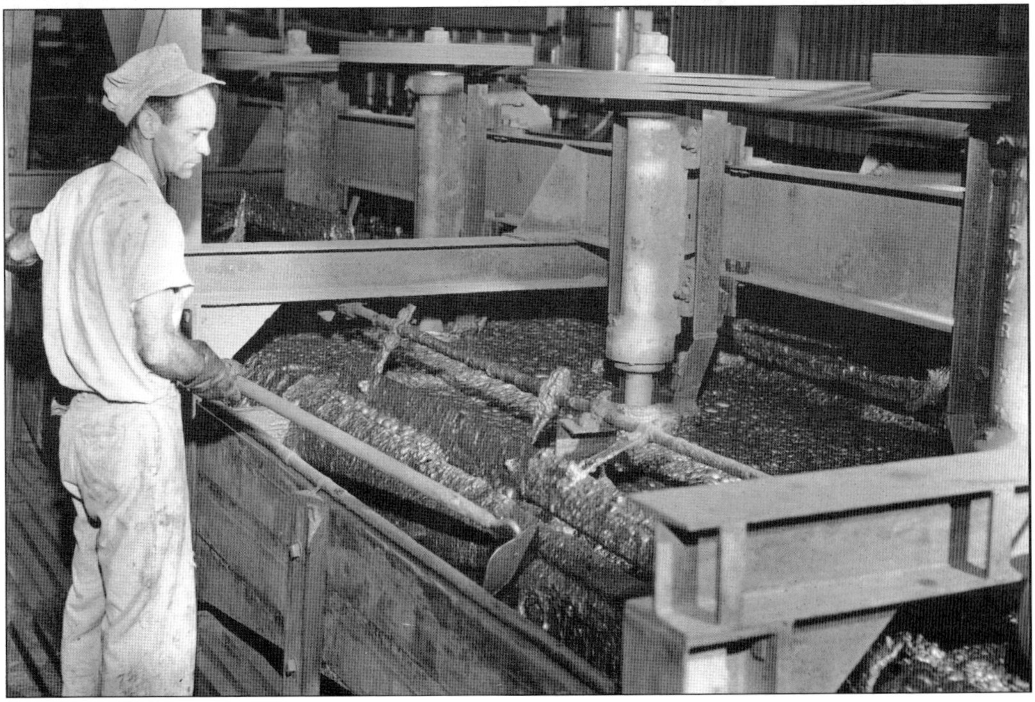

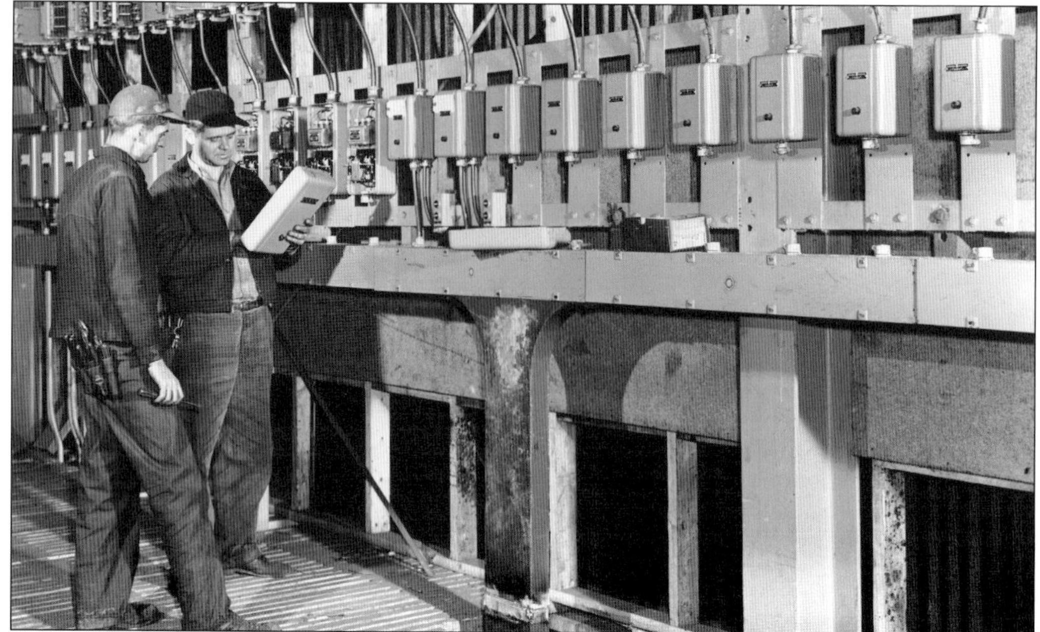

Electricians, 1954. The TCC power and electric division was responsible for every aspect of the company's electrical systems, from motor maintenance to line work, transformer installation, substation inspection, switching, and installation of new systems. Above, W.E. Hardeman and Walt Walden work on installing a new bank of control boxes at the Isabella flotation mill. Below, Gene Chancey checks the commutator brushes on the generator of one of the company's diesel locomotives. This division was also responsible for the maintenance of crossing signals, shop lighting, and any other electrical chores needed by the railroad department.

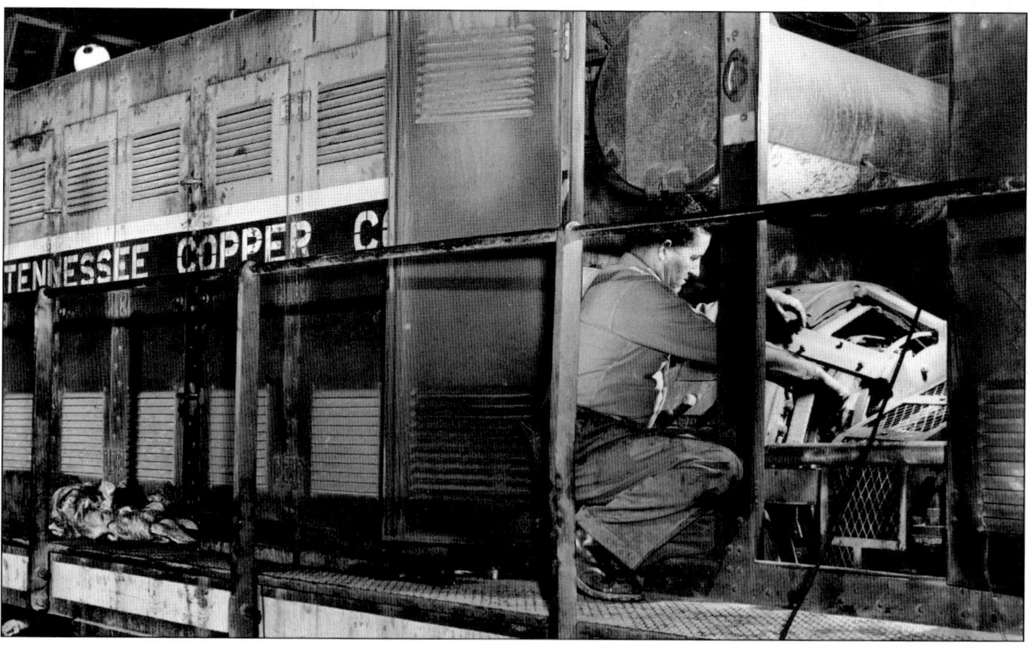

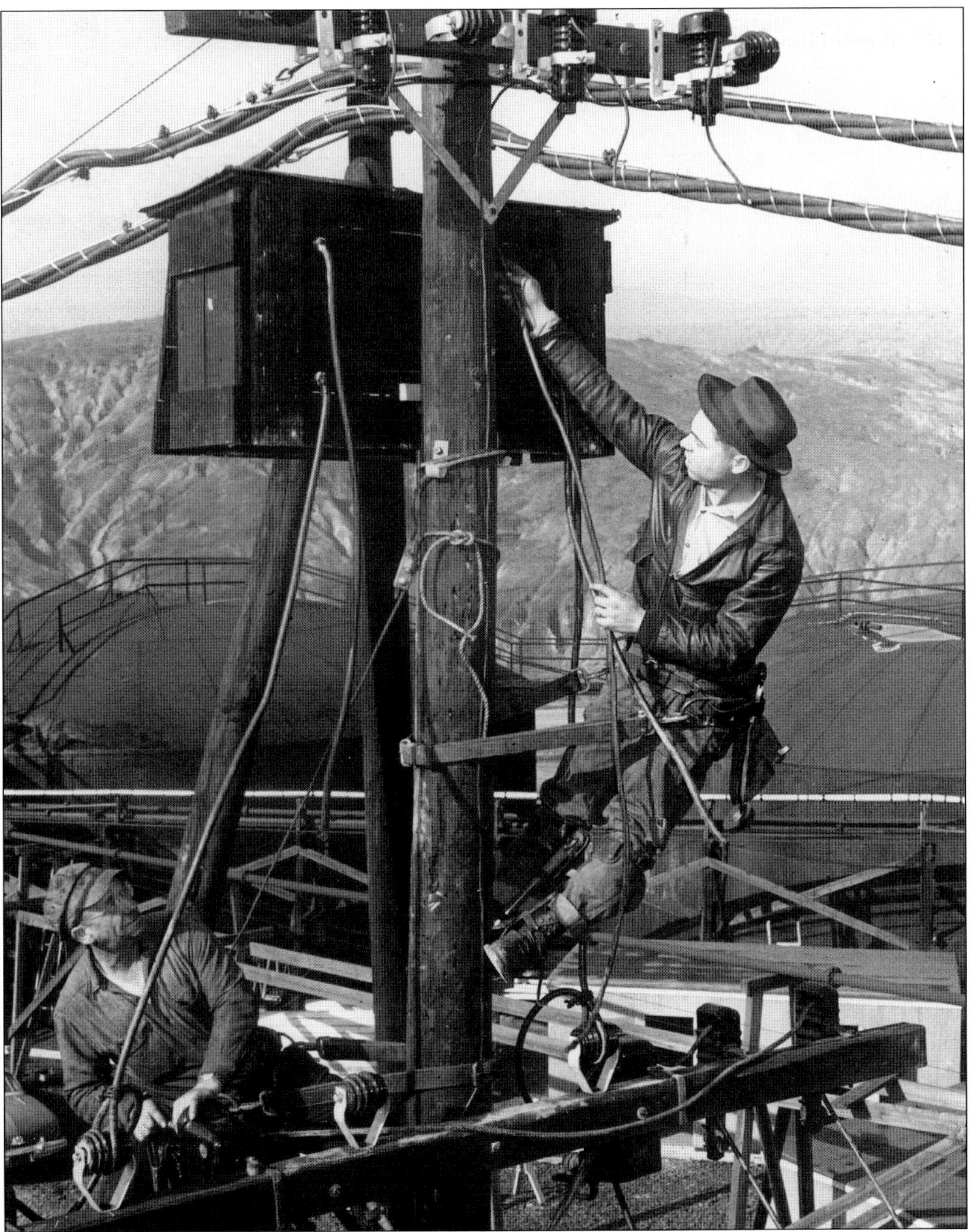

Electrical Upgrades, 1954. After tying in a new 2,300-volt feeder line, electricians Fred McCay (left) and Clement Bailey connect the new line to the transformer bank at the other end. The feeder line was part of a major project to rebuild the electrical distribution system in the entire TCC plant at Copperhill. McCay retired in 1965 after 24 years as a TCC and Cities Service electrician; Bailey, a World War II veteran who had joined the company prior to the war, was working as an electrician in the Cities Service railroad department when he celebrated his 40th company safety anniversary in 1981.

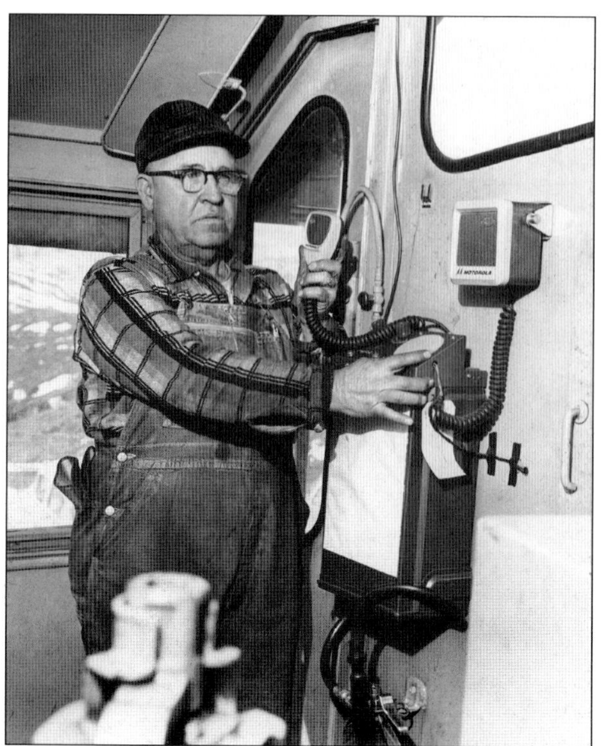

Railroad Department, 1958. Conductor W.C. "Chubby" Smith demonstrates the use of the new radiotelephone system that simplified communications within the TCC railroad department. Prior to the installation of the new system, Smith would have needed to leave the train, go to a land-based telephone, call the office for clearance to proceed into the station with a load of ore, then return to the train and convey the orders to the engineer. Below, railroad chief clerk Tyson Smith uses the new system to give orders to a train crew, while assistant railroad superintendent Cecil Kell (center) talks with brakeman Hoyt Jones through the open window.

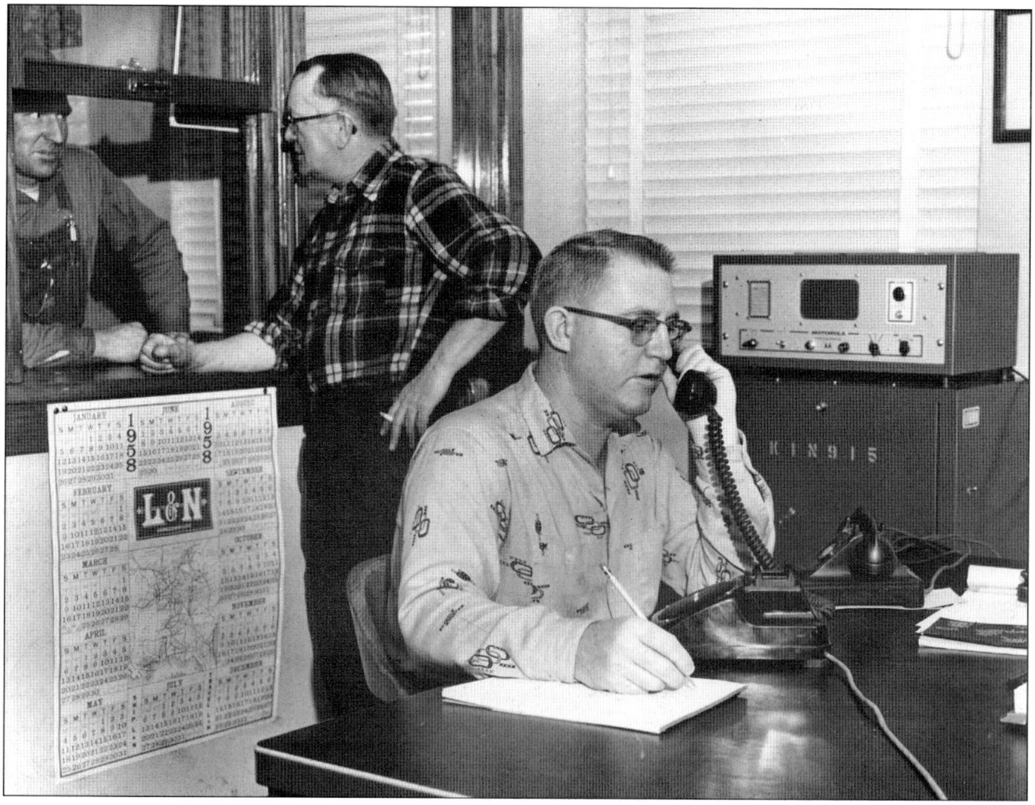

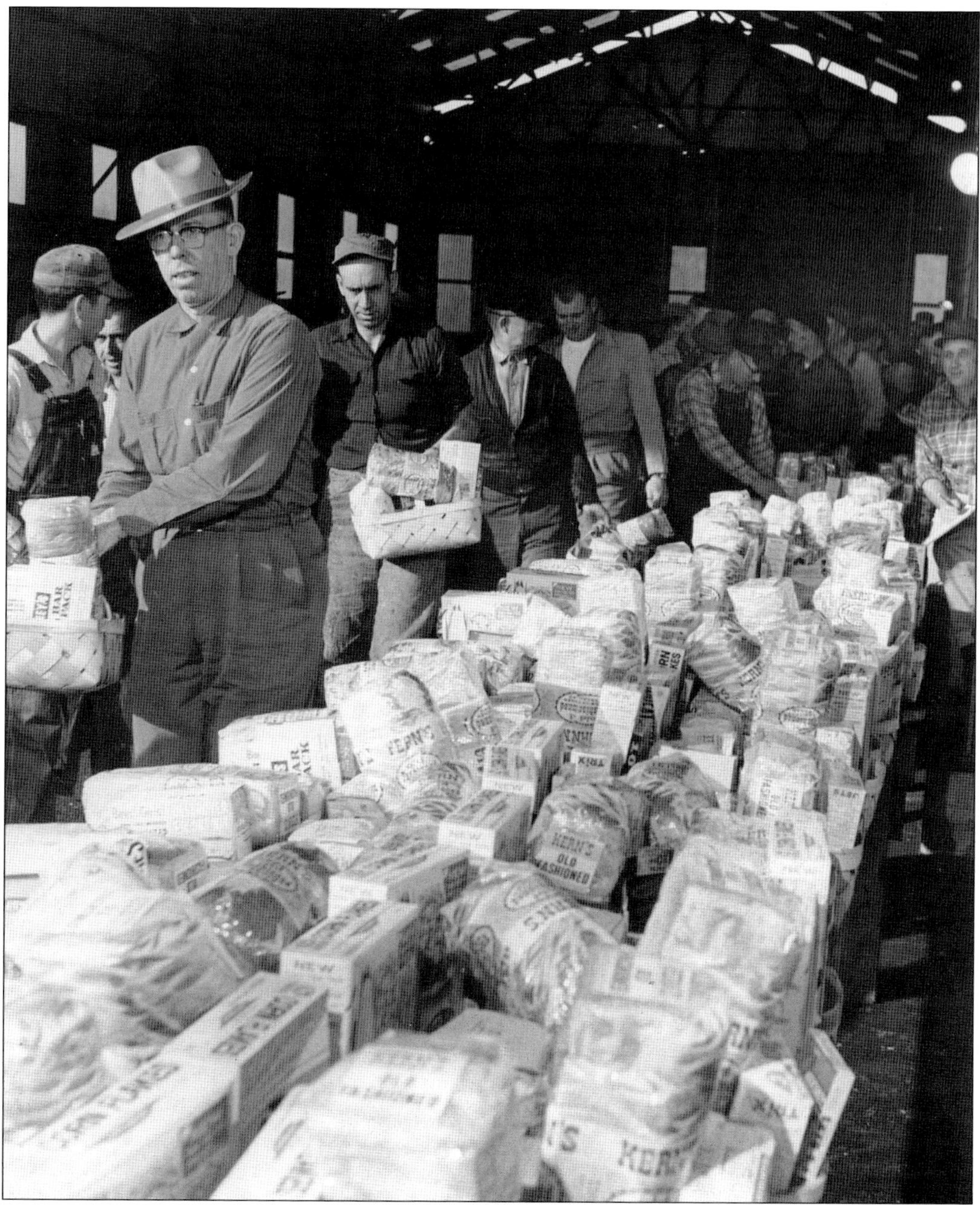

Safety Recognition, 1959. The railroad department celebrates two accident-free years with food-filled safety award baskets for department employees. Picking up their baskets, from left to right, are H.M. "Gid" Ware, Clifton Williams, Richard Long, Vince Williamson, and Otis Adams. At far left is brakeman L.M. "Buster" Dillard, and at far right, on the other side of the table, is Carlton Williams. Prominent in the baskets are loaves of Kern's Old-Fashioned Bread, a local favorite from a nearby Knoxville, Tennessee, bakery. The railroad department had just been cited for its safety record of 832 days, about 466,000 man-hours, without a disabling injury.

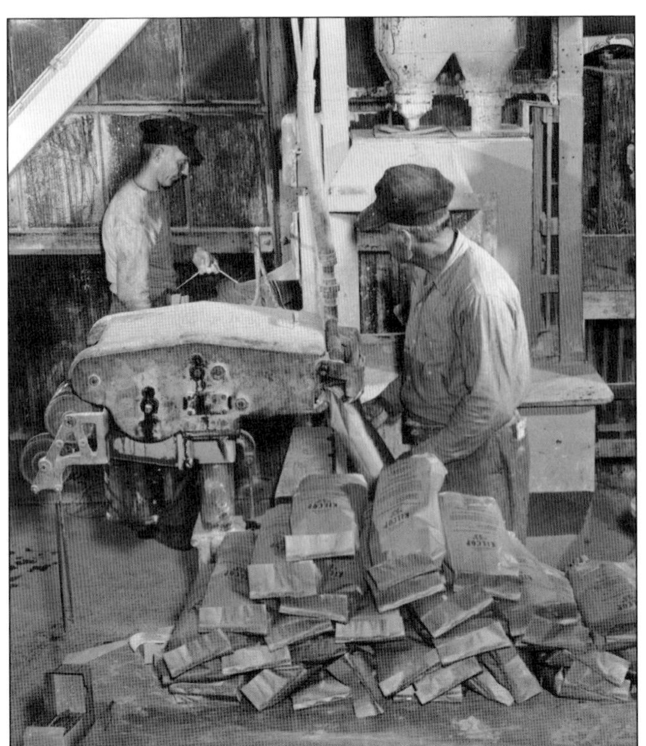

Copper Sulfate Workers, 1953. Copper sulfate was first produced at TCC in 1922 and was in demand as a primary ingredient in fungicides. By the time this photograph of a bagging operation was taken in late 1953, the company's product line included custom fungicides. The small bags are labeled "Kilcop 53," a brand name that had been trademarked the previous year by the Kilgore Seed Company.

Fan Repair, 1953. Using brazing rods, veteran lead burner Cliff Cunningham (right) and an unidentified coworker apply a lead-zinc alloy to one of the metal fans that moved gases through flue ducts at the Copperhill acid plant. The fans were subject to corrosion from the gases and needed periodic repair.

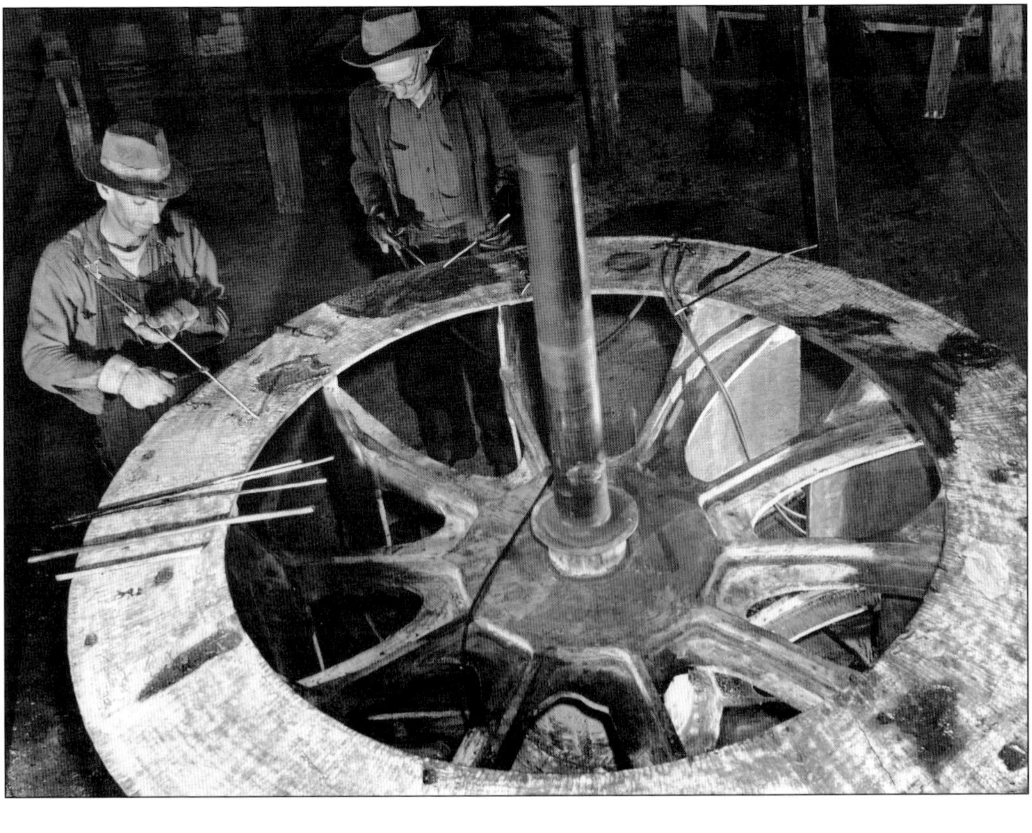

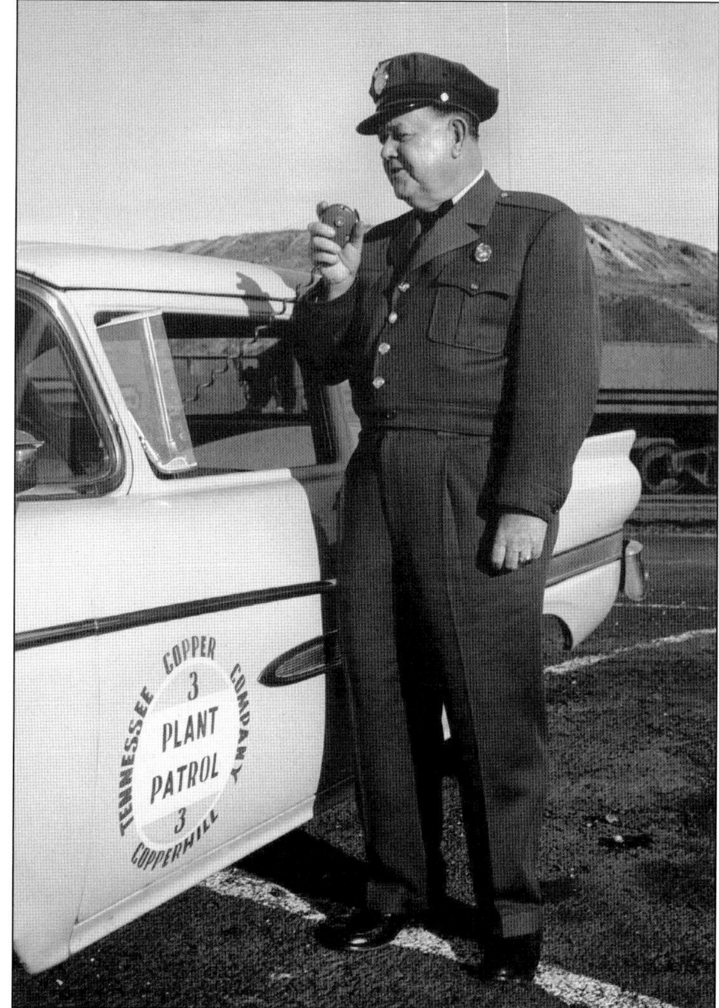

GUARD FORCE UPDATES, 1959. Patrolman Clint Ingle shows off the guard force's new, tailored forest green whipcord uniforms and demonstrates the new radio communication system that had been installed during the summer. The system included a base unit in the gatehouse and transmitters with small microphones in each of the two plant patrol cars.

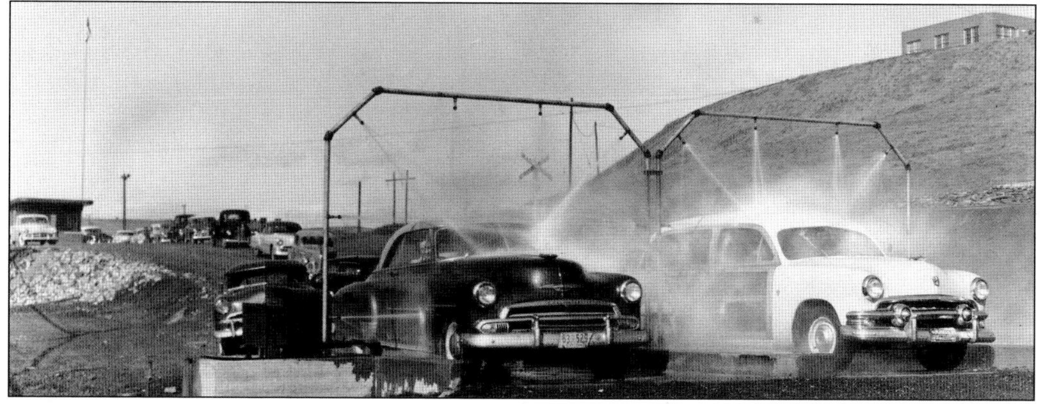

AUTOMATIC CAR WASH, 1954. In spring 1954, TCC installed a new employee amenity: a two-lane automatic car wash just outside the main gate at its Copperhill plant. Its popularity, and perhaps its necessity after a day in the industrial plant environment, is evident from the long line of cars waiting their turn at the end of a shift.

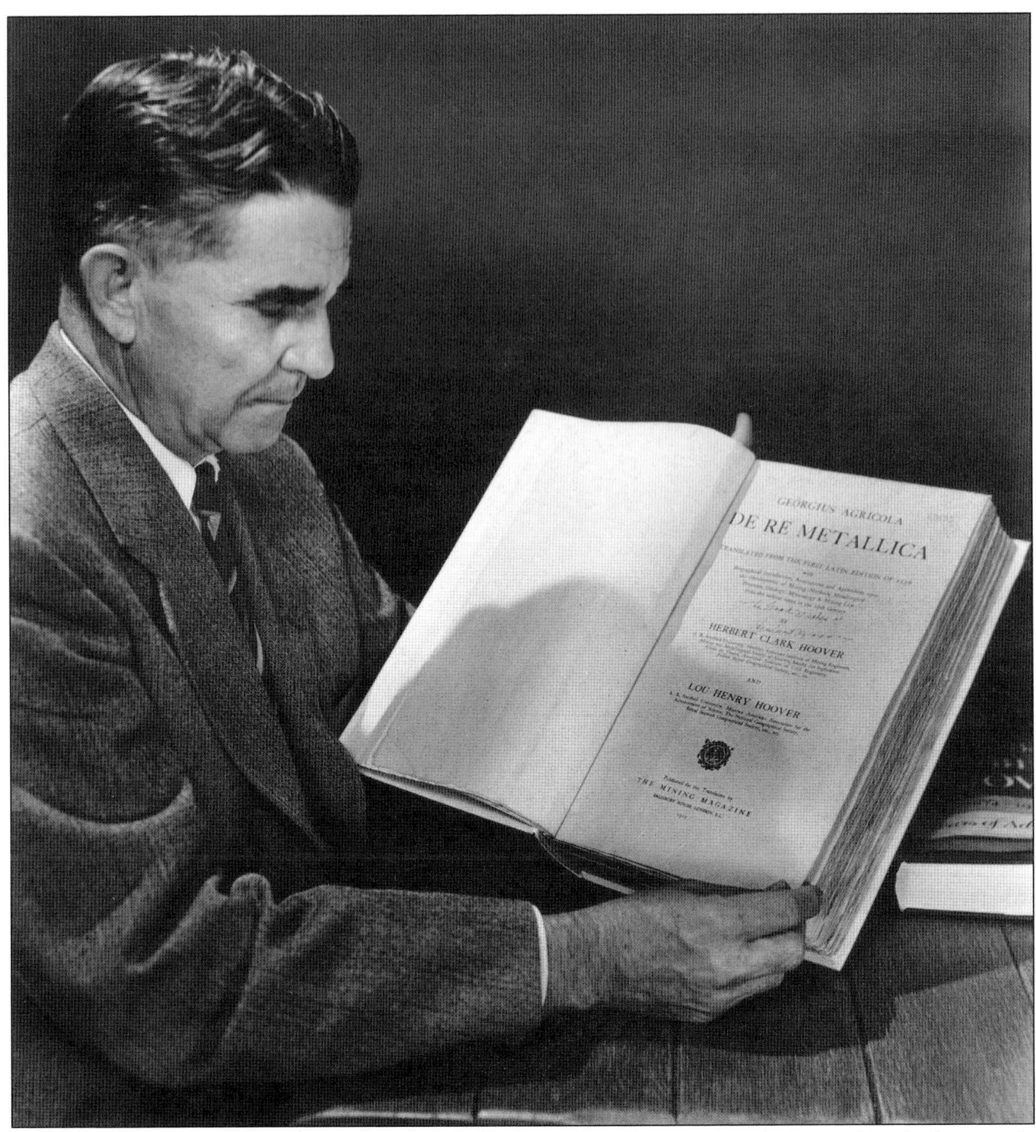

CHANGE AND CONTINUITY. The mining and processing of copper can be traced back thousands of years. But it was the invention of the printing press in the middle of the 15th century that made it possible to pass along the accumulated knowledge of those years in written form. Among the prized possessions of Edward M. Jones, superintendent of the TCC chemical engineering department, was a rare English translation of the 1556 book *De Re Metallica*, an extensive treatise on the mining methods and processes of the Renaissance era in Europe. The first English edition had been translated and published in 1912 by mining engineer and future US president Herbert Hoover and his wife, geologist Lou Henry Hoover. By the time Jones acquired his autographed copy of the original Hoover translation in 1952, he had worked at TCC for 30 years and was credited with several innovations and improvements to the company's own processes. The rare book, one of just 3,000 copies printed, is now in the permanent collection of the Ducktown Basin Museum.

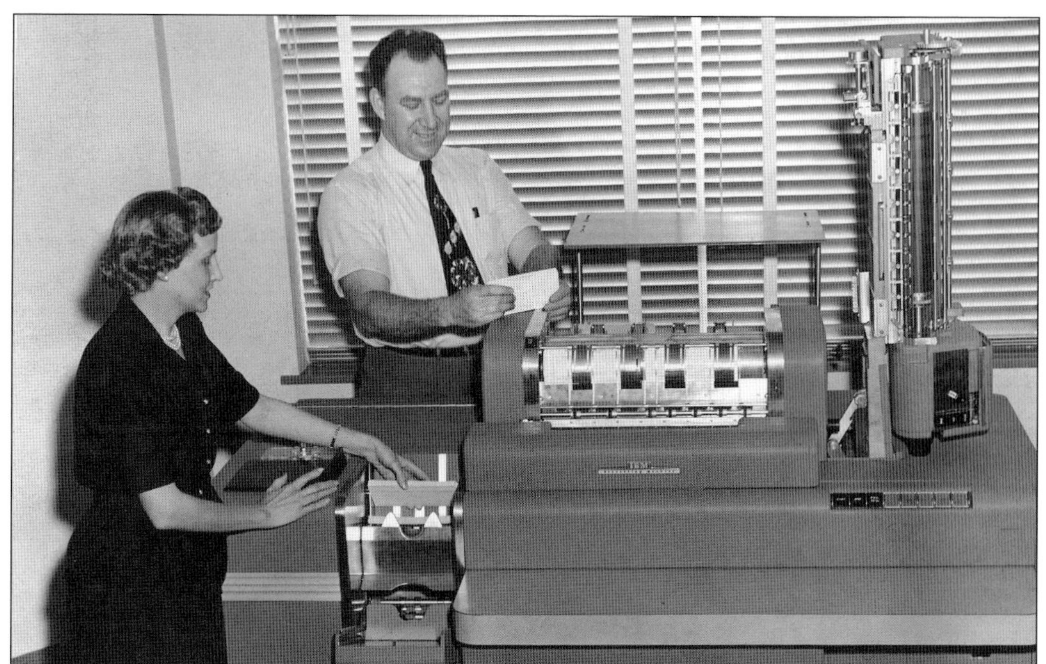

Payroll Department, Early 1950s. Above, Margaret Duncan and Bob Beckler pose with the new IBM accounting machine that heralded the automation age in the TCC payroll department. The machine could print out 30 checks a minute. Employees were cautioned not to fold the checks, which were printed on thin card stock and had to be fed back through the machine when they were returned by the bank. Below, chief clerk Robert E. Barclay (center), Helen Epperson, and Julius Taylor inspect the first TCC payroll book, which is dated April 14, 1899. The payroll book recorded that a total of 50 men were employed at TCC by the end of April, most of them at a wage of $1 a day, with an occasional skilled worker earning $2 a day.

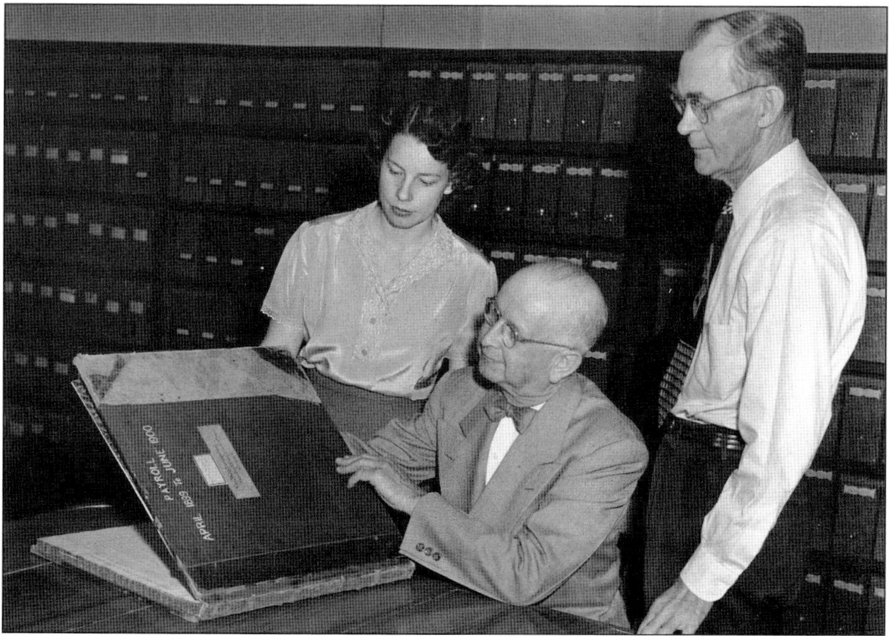

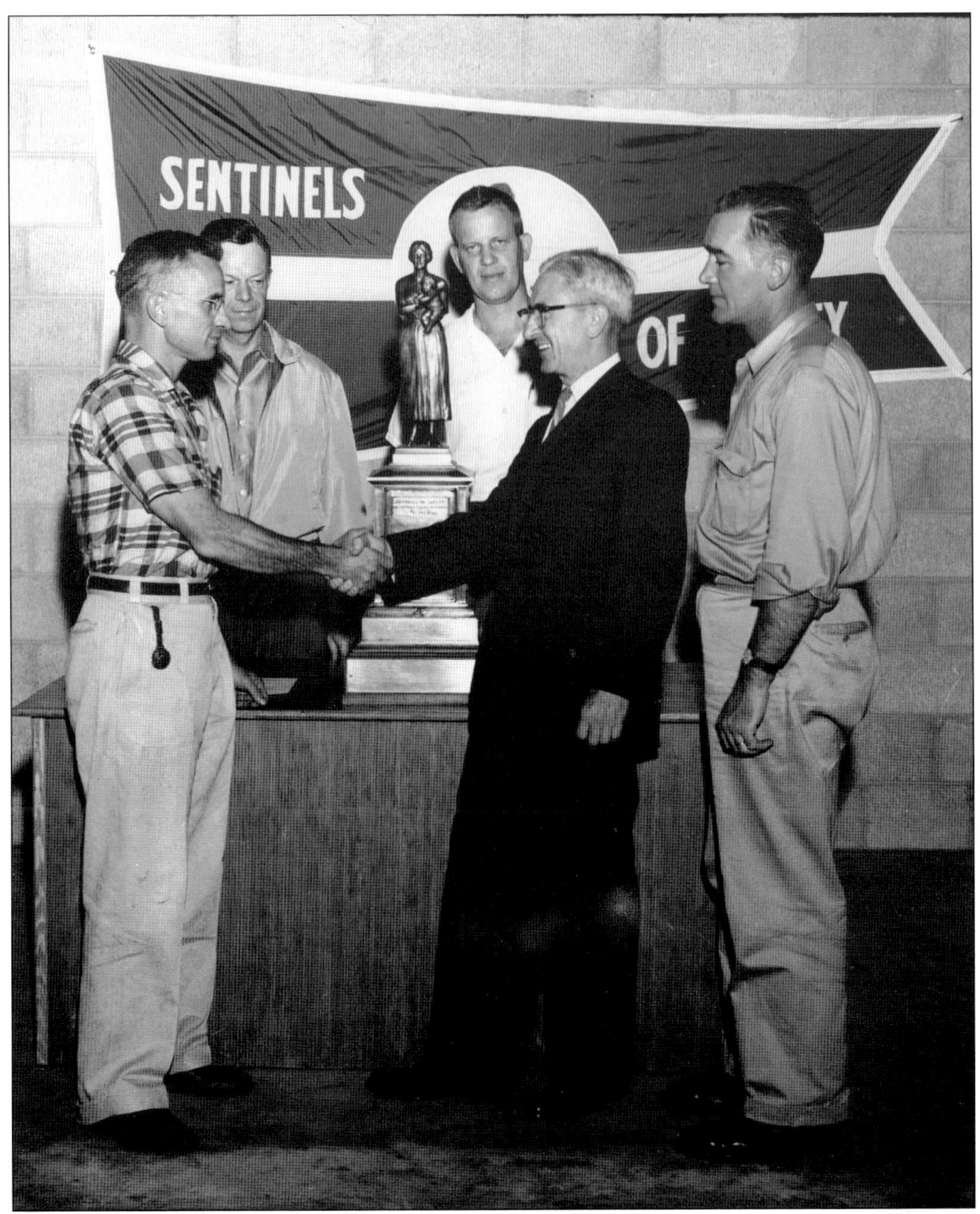

Sentinels of Safety Award, 1958. In 1957, in the US Bureau of Mines national safety competition, TCC's Calloway and Mary Mines were jointly named the safest underground metal mine workings in the United States. TCC had won the coveted Sentinels of Safety award on four previous occasions: at Burra Burra Mine in 1933 and 1946, at Boyd Mine in 1942, and at Calloway-Mary in 1953. The 1957 win set a national record for TCC, as no other mining company in the United States had ever won the award five times. In this October 1958 photograph, Calloway-Mary mine foreman H.B. Estabrooks (left) accepts the bronze statuette trophy from Thomas E. Miller, contributing editor to the *Explosives Engineer*. The others, from left to right, are mine superintendent H.F. Kendall, safety engineer A.D. Annand, and Calloway-Mary shift foreman Harold Dale.

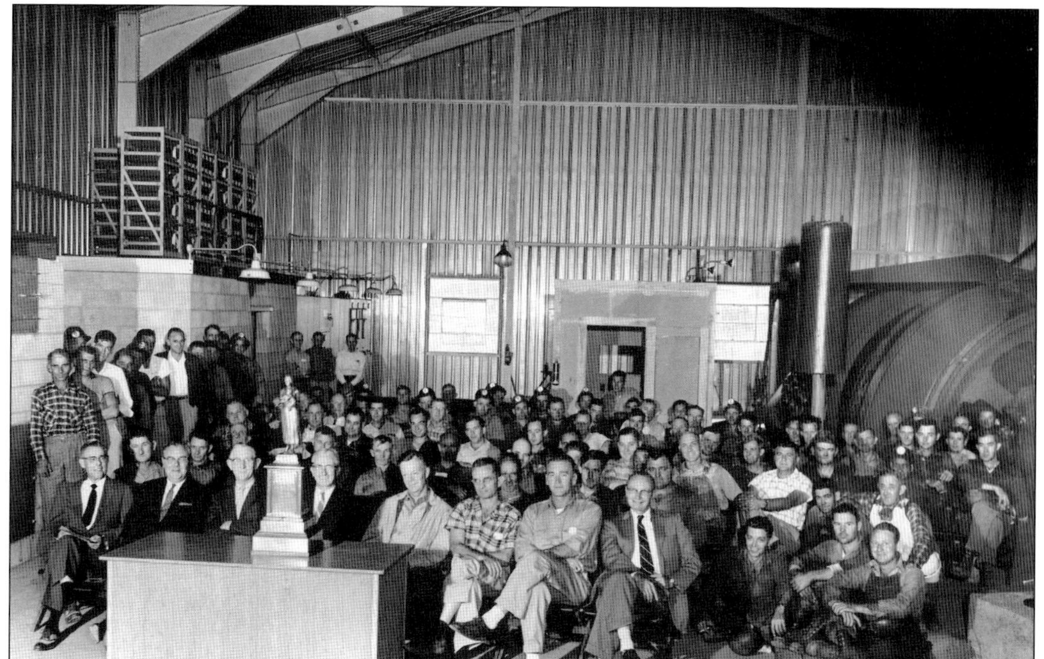

Sentinels of Safety Ceremony, 1958. In October 1958, the supervisors and crew at Calloway and Mary Mines gathered for the official presentation of the 1957 Sentinels of Safety award from the US Bureau of Mines. The ceremony was held at the new Calloway "B" shaft hoist house, which was still unfinished. Calloway Mine won the award in 1966 for an unprecedented third time.

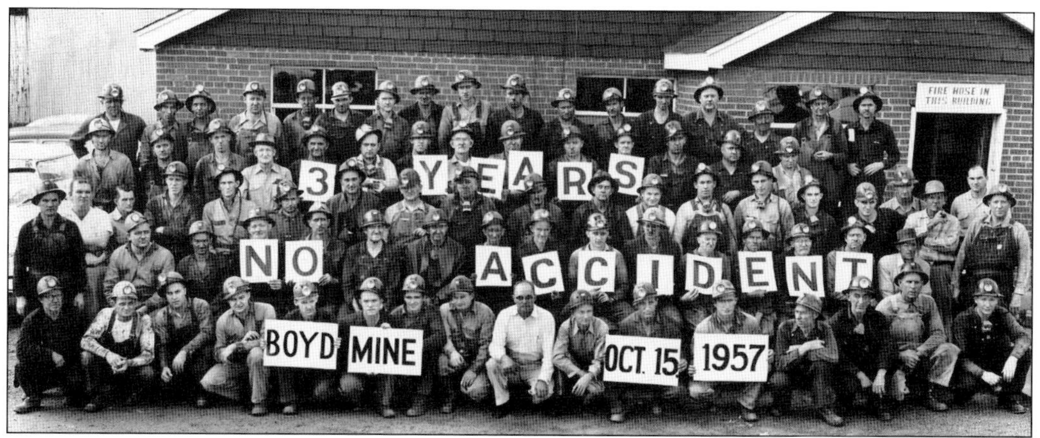

Boyd Mine Safety Celebration, 1957. Calloway and Mary miners apparently had some stiff competition from within TCC itself. On October 14, 1957, Boyd Mine set a new TCC mine-safety record: three years without a lost-time accident. Of the 88 members of the Boyd crew, 65 had worked there for the entire three-year period. Kneeling at center in the first row, in the light-colored shirt, is mine foreman Gus Nelson.

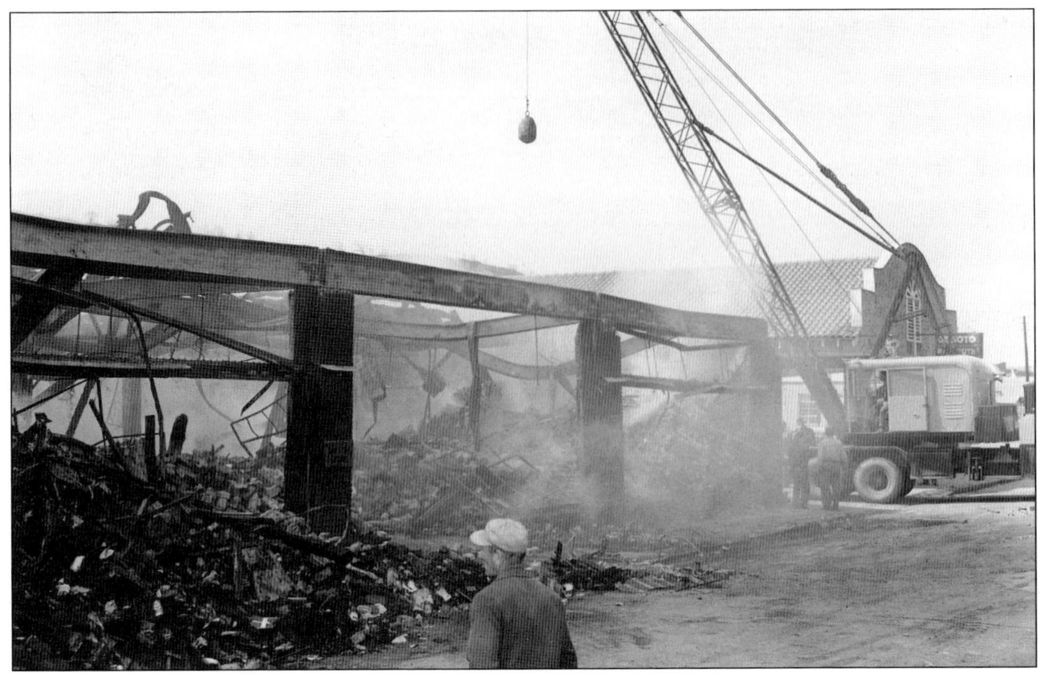

Smelter Store Fire, 1958. On the night of March 22, 1958, TCC's venerable Smelter Store in downtown Copperhill was destroyed by fire. It was still smoking the next morning when company employees arrived to start clearing the street, using a crane to knock down the remains of the brick walls. Soon afterward, TCC engineering teams went to work to design a new store building in-house, with multiple departments contributing their individual expertise to the project. Below, from left to right, electricians Glenn Harper and Fred McCay, electrician helper Calvin Cook, electrician Raymond Arp, and project foreman Joe Quintrell prepare to tackle a complex six-conduits point where the new store's telephone system would go underground.

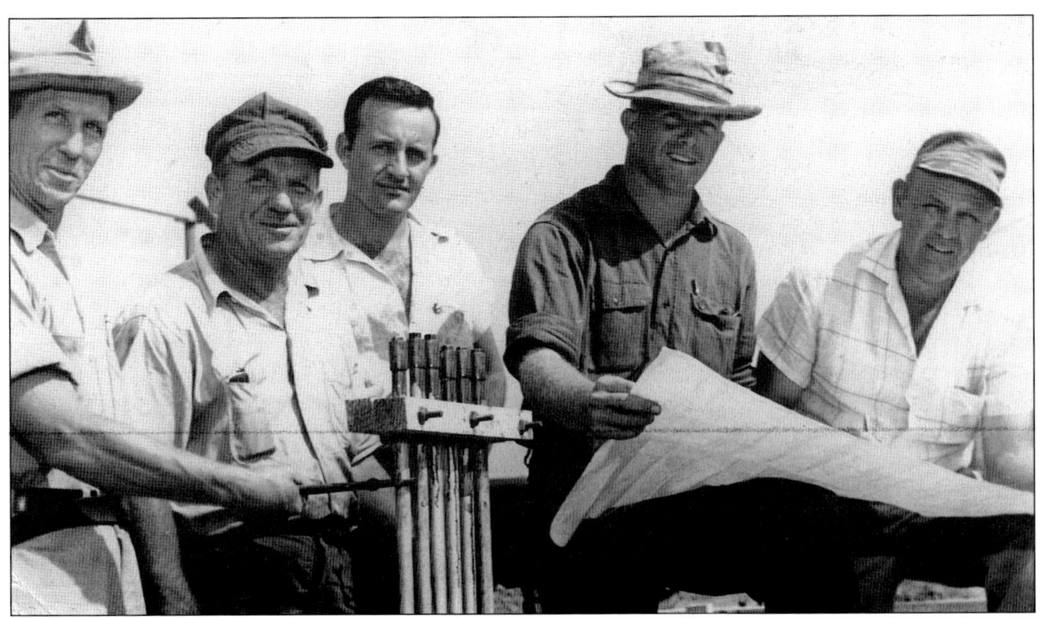

SMELTER STORE CONSTRUCTION, 1958. When construction on the new store building began in August, teams of company carpenters, stonemasons, bricklayers, pipefitters, electricians, and members of other specialized trades spent four months bringing the engineering department's blueprints to life. In this photograph taken during construction, general carpenter foreman Frank G. Beaver overlooks the work in progress.

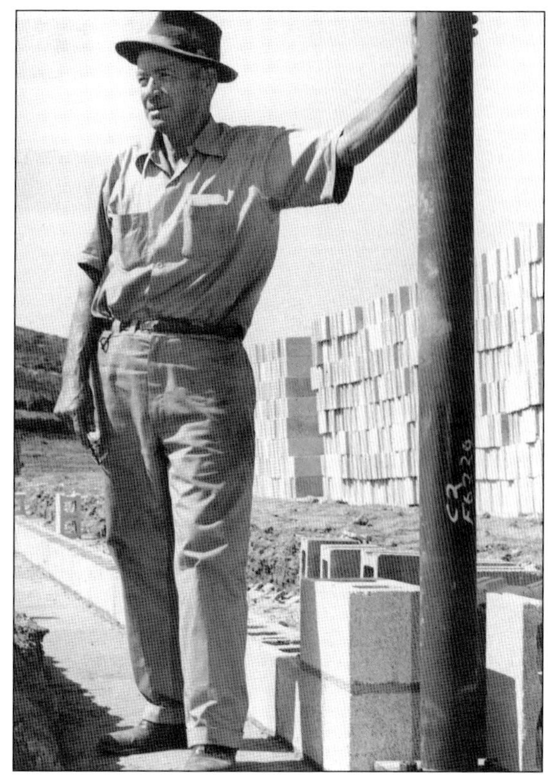

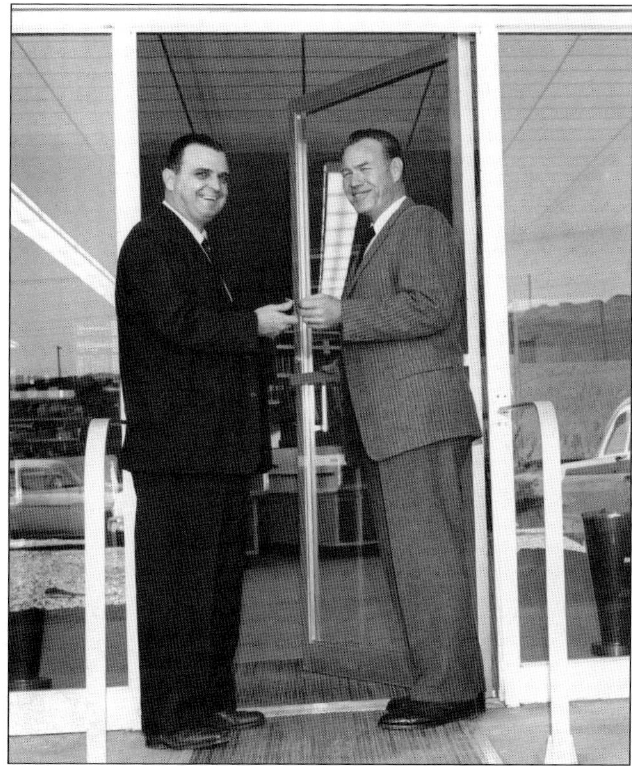

SMELTER STORE OPENING, 1958. Just four months after the beginning of construction, general construction foreman Carl Crockett (right) hands over the keys of the new 35,700-square-foot building to store manager W.E. Roach. The new store held a community open house on November 30, 1958, and officially opened for business on December 1.

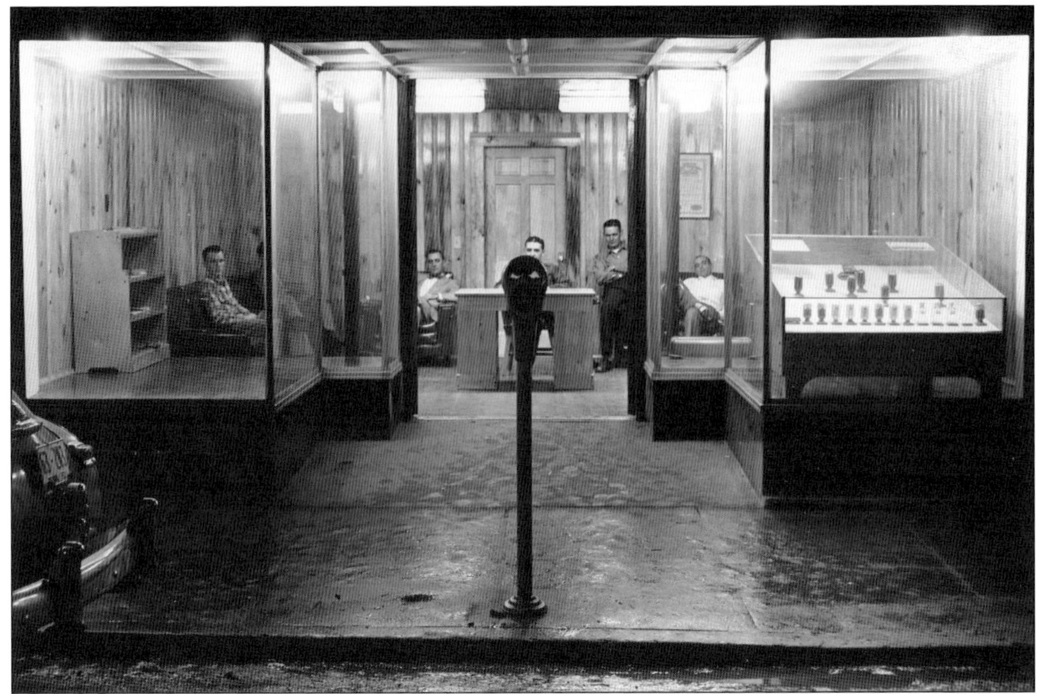

New Union Headquarters, 1953. Above, members of International Chemical Workers Union (ICWU) Local 401 enjoy the comforts of their new headquarters in downtown Copperhill. From left to right are Alvin Kyle, union trustee W.J. "Junior" Sparks, union financial secretary Kenneth Collins, Frank G. Fox, and Joe Chancey. Below, from left to right, Sparks, union president Earl G. York, Fox, Kyle, and Chancey pose with a window display on loan from TCC. Members of the union, who represented a variety of skills and crafts, had done all the renovation work free of charge in their off-hours.

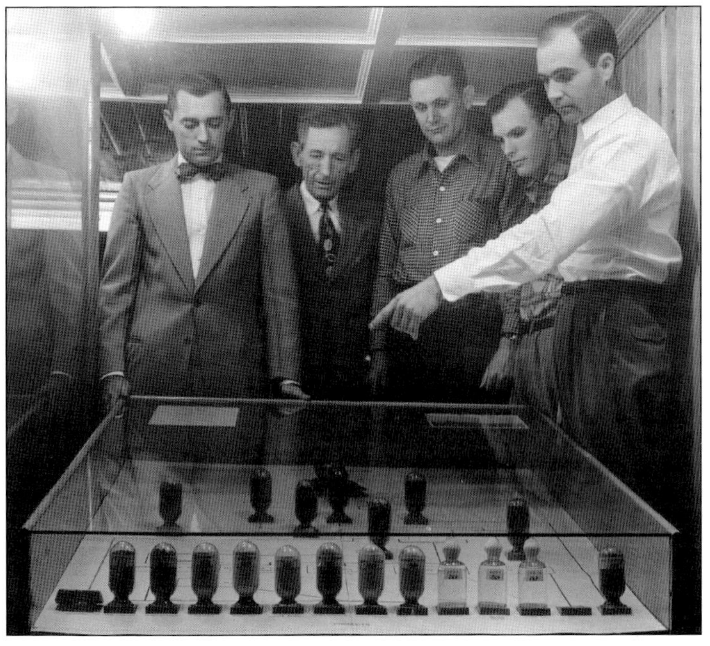

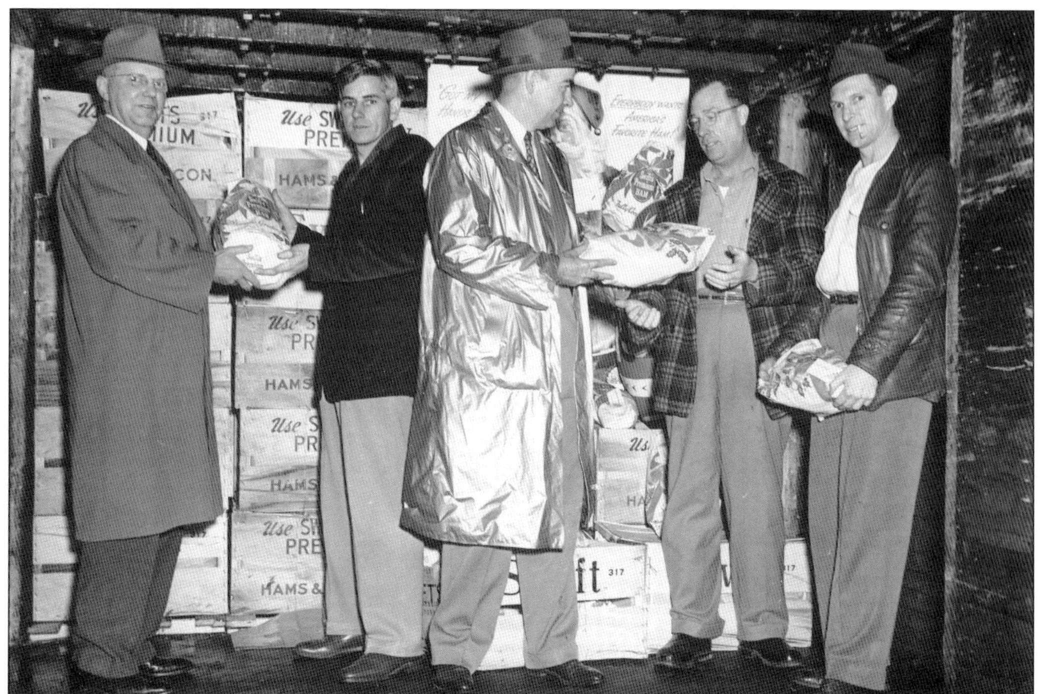

Hams for the Holidays, 1951. TCC general manager T.A. Mitchell (left) and assistant manager R.R. Burns (center) present newly arrived Swift hams to union officials. In 1951, the company also began a tradition of giving holiday hams to retired employees.

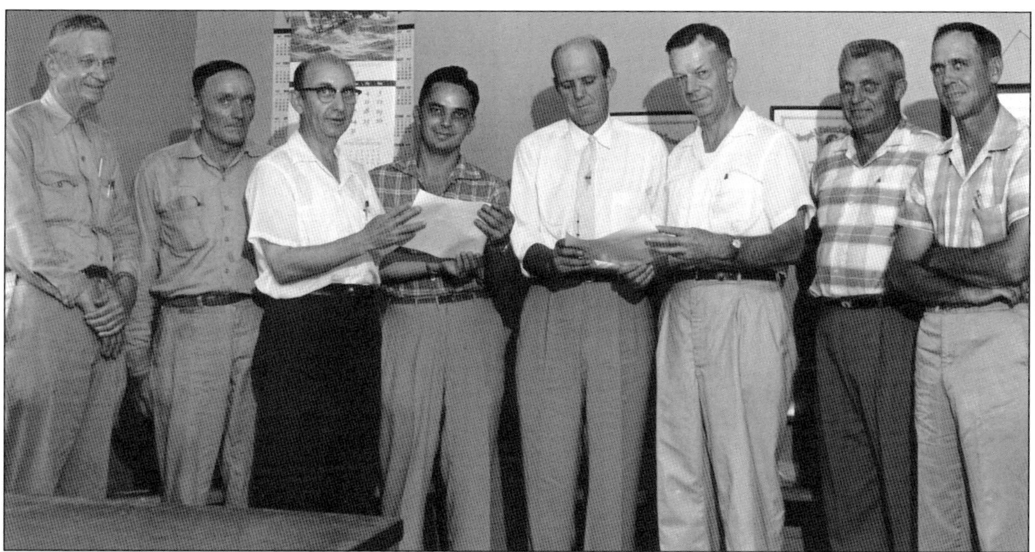

ICS Graduates, 1958. In 1954, in coordination with International Correspondence Schools, TCC instituted a voluntary home-based employee training program intended to augment the company's on-the-job training. Electrician Bazelle Nelson and carpenter Ray Williams were among the first graduates. From left to right in this presentation of completion certificates are Joe Lanning, Lester Smith, assistant mine superintendent Roy Clay, Nelson, Williams, mine superintendent H.F. Kendall, Creed Stillwell, and J.D. Galloway.

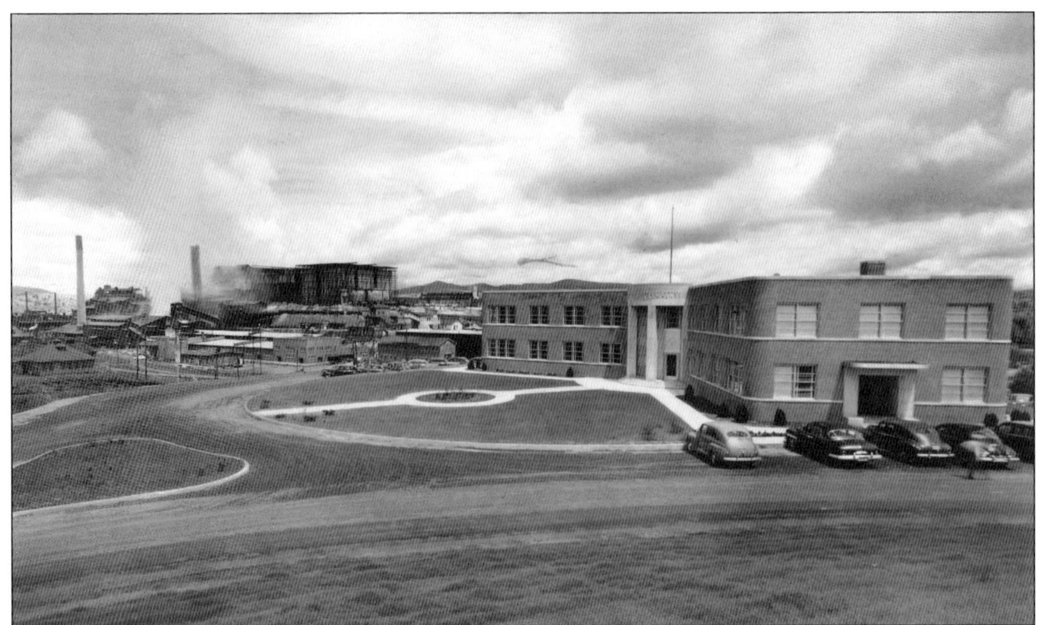

TCC BY DAY, 1950S. The streamlined design and manicured grounds of the new General Office stand in stark contrast to the adjoining plant site, which had been designed strictly for its industrial functions. Judging from the automobiles parked by the side entrance of the office building, this photograph appears to date from the early 1950s.

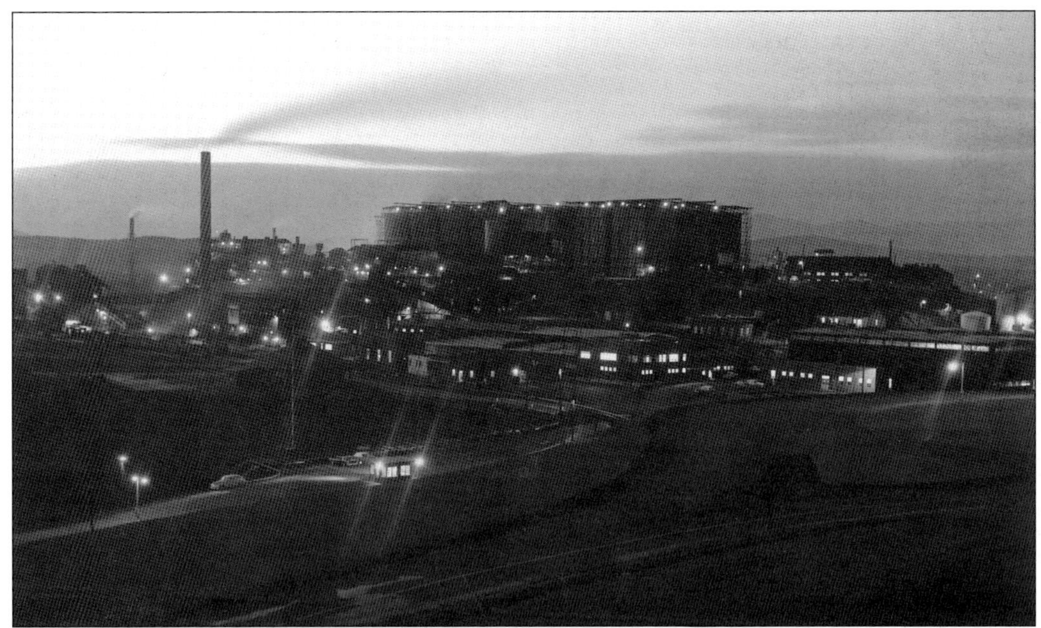

TCC BY NIGHT, 1950S. At night, with every light blazing against a darkened sky, the Copperhill plant took on the aspect of a small city. Within a few years, the No. 1 chamber acid plant that dominates the skyline in this 1959 photograph would be demolished as the company converted exclusively to the contact acid process.

Three

The 1960s

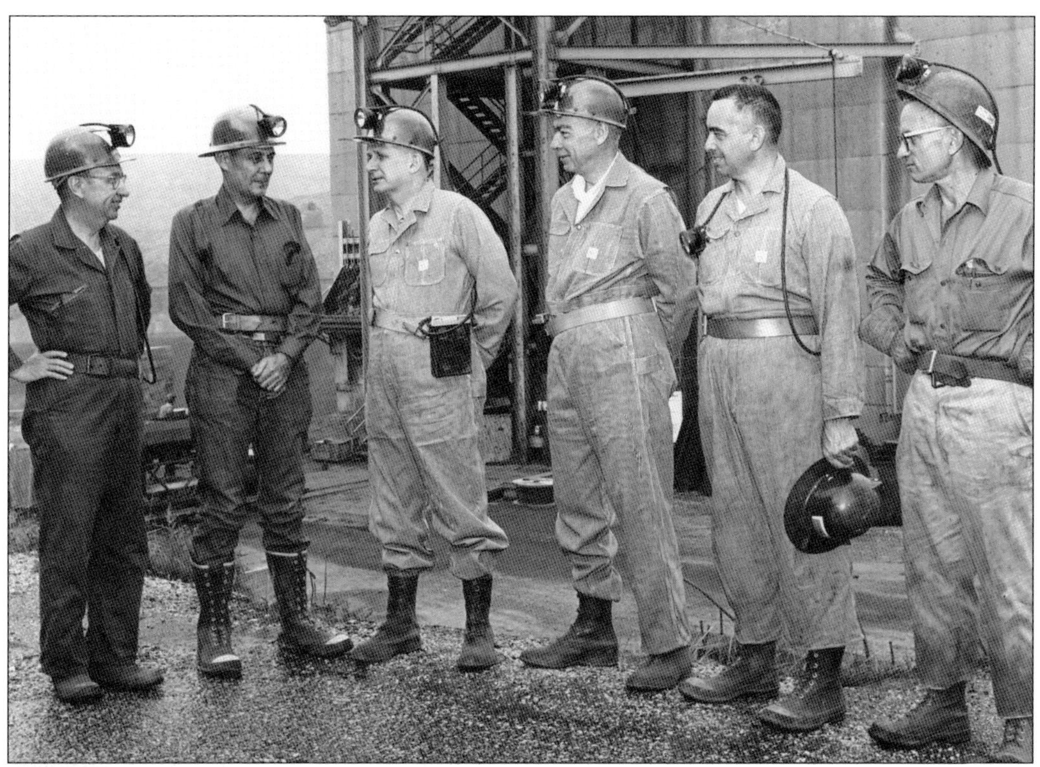

Management Tour, 1966. The 1960s, particularly the period after the company's acquisition by Cities Service Company in 1963, was a decade of change for TCC. From left to right, mine superintendent Roy G. Clay, assistant general manager W.F. Keffer, Cities Service president J.E. Heston, Tennessee Corporation president R.R. Burns, chief geologist Maurice Magee, and assistant mining superintendent H.B. Estabrooks prepare for an underground tour of Calloway Mine.

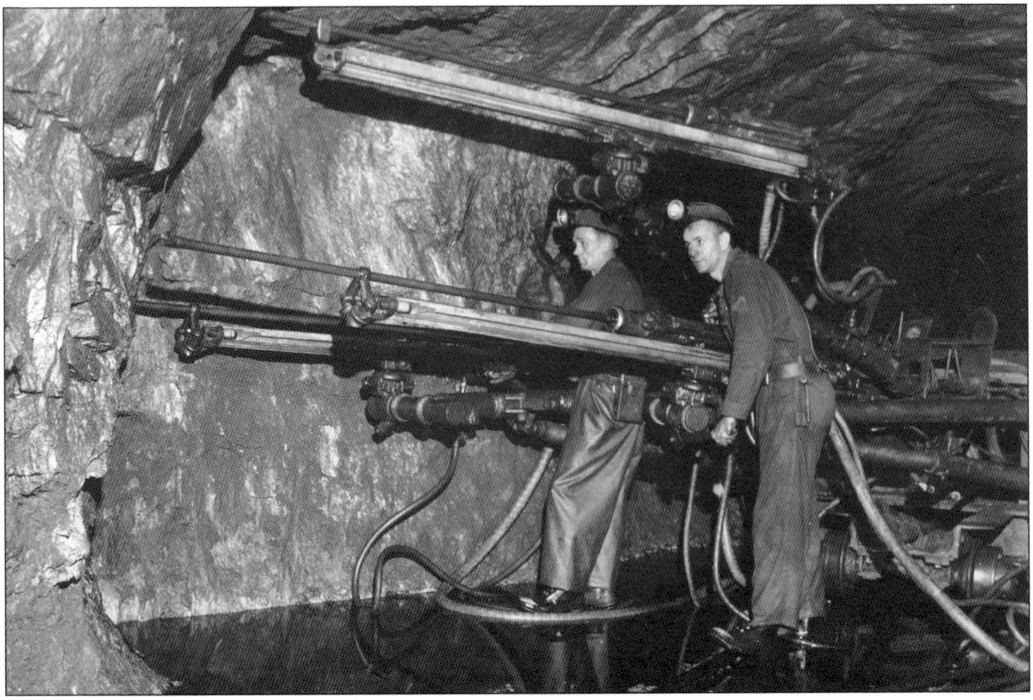

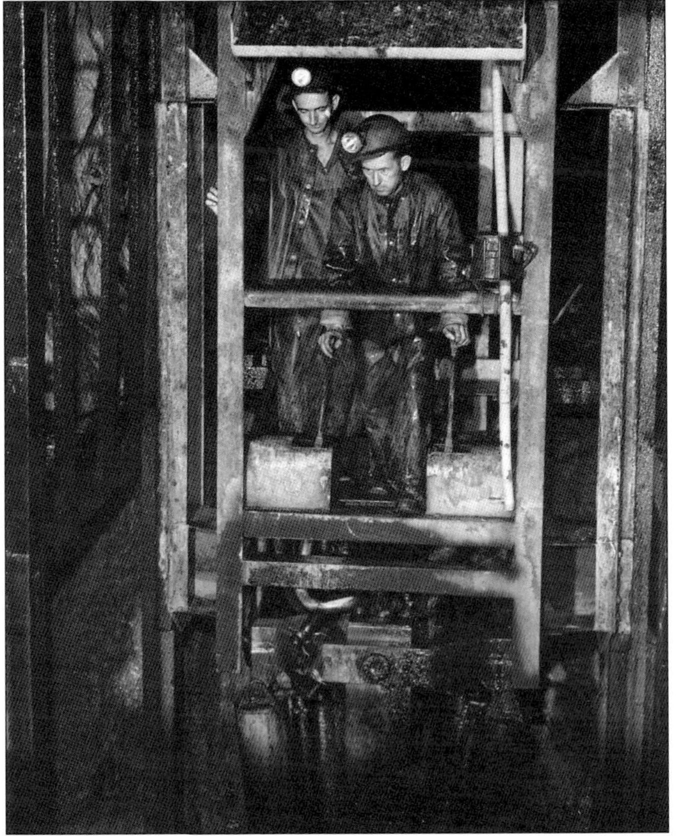

Developing Cherokee Mine, 1960. On August 23, 1960, after two years of work, TCC drillers completed the driving of a tunnel from Cherokee Drift, which joined Boyd Mine at Central Shaft, to a short crosscut tunnel from the recently sunk Cherokee Shaft. The new tunnel was more than 6,000 feet long and was the last phase in opening up access to the ore body at Cherokee Mine. Above, drillers W.K. Stover (left) and Almer Roberts position a hydraulically operated jumbo in preparation for creating a lay-by where mine cars could be shifted during further advancement of the drift. At right, C.W. Falls (at the controls) and E.E. York operate a Cryderman machine, a remote-control method for removing blasted rock from the bottom of the shaft to the surface.

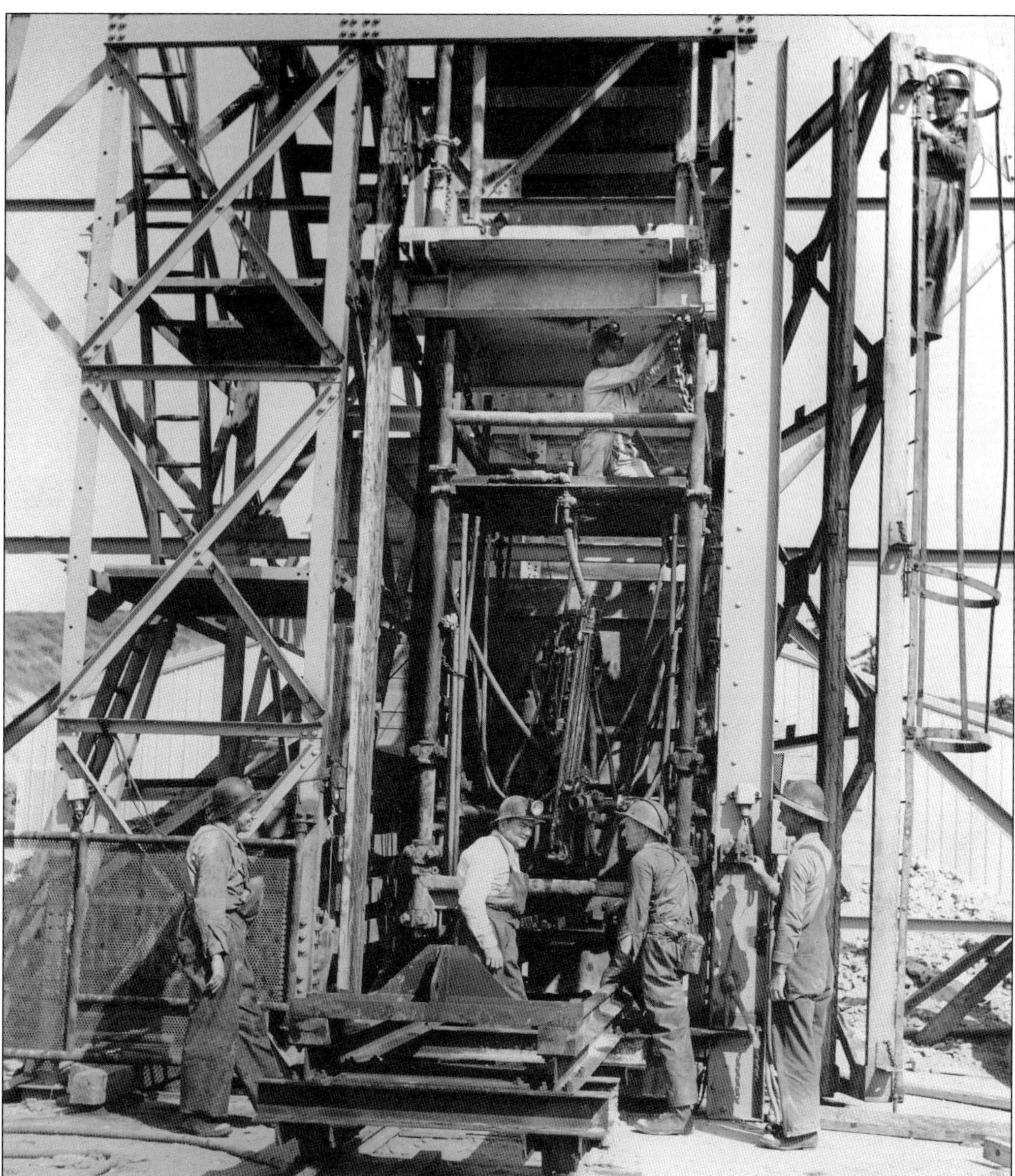

Drilling Jumbo at Cherokee Mine, 1960. In June 1959, while drillers were driving the original Cherokee Mine drift connecting the mine site to Central Shaft, another drilling crew began the sinking of the mine's main shaft, which would open an ore body that had never been explored fully and had been closed for nearly a century. In this photograph, published in September 1960, drillers prepare to lower a five-drill jumbo to the bottom of the shaft by hooking it to a work cage. Kneeling on top of the jumbo and hooking the chain is C.E. Bennie. Standing at the top of a hinged section of the headframe, which had been swung out for placement of the work cage, is Fred Garren. Standing at the shaft collar, from left to right, are Frank Callihan, O.W. Rucker, E.L. Jones, and shaft drill foreman H.A. Pickelsimer.

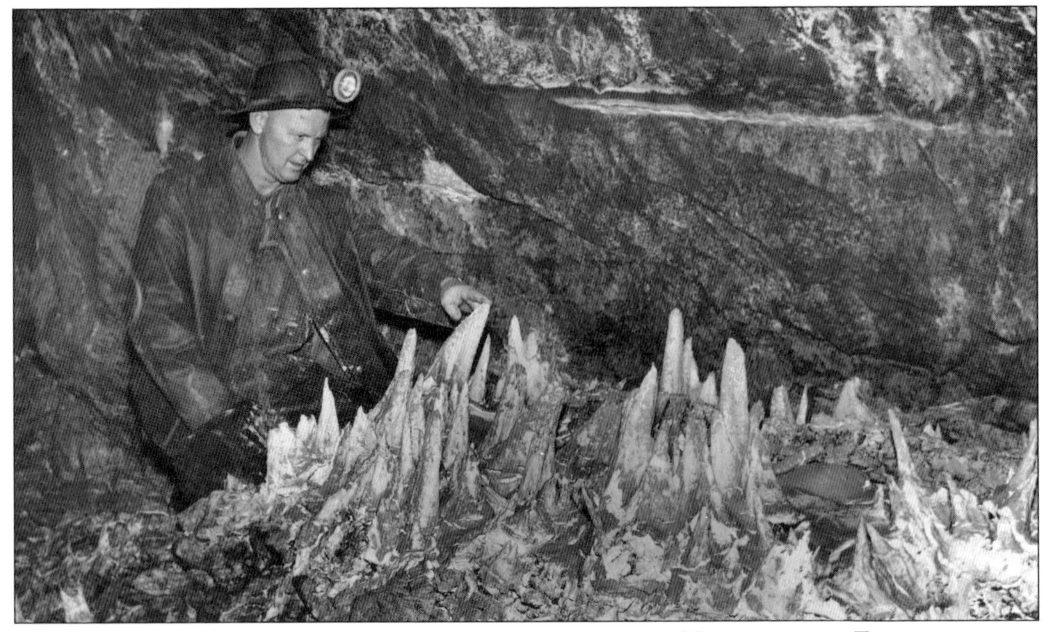

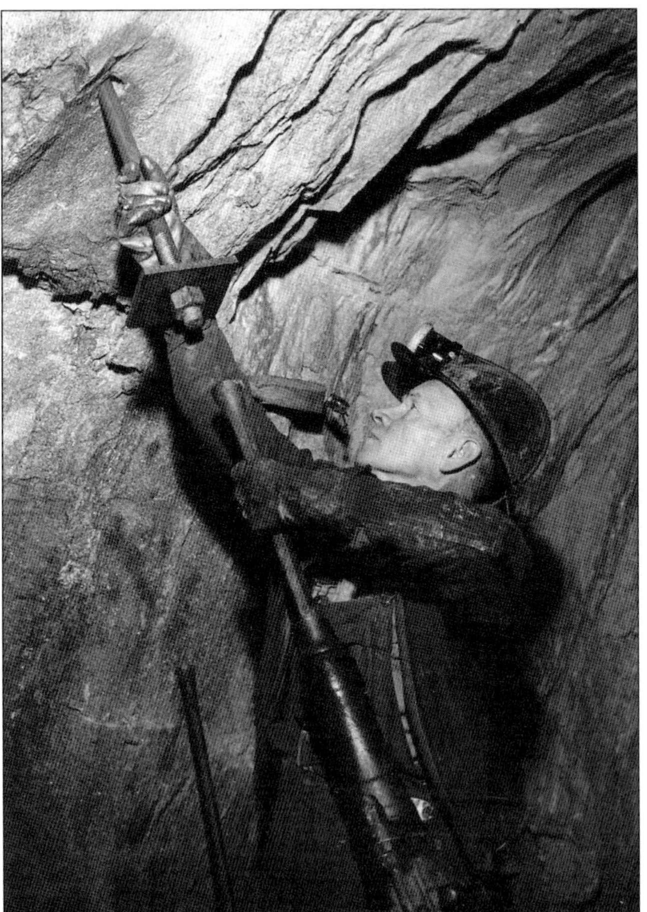

Underground Exploration, 1961. During the development of the Cherokee Mine property, the shallow former DC&I mine known as the Westervelt shaft was reopened after having been flooded for almost 40 years. Among the curiosities found by timberman R.C. Spurling and his co-workers when they reached the first level were these porous mineral deposits that had grown into a shape resembling small stalagmites.

Rock Bolting, 1961. Wade Anderson prepares to drive a six-foot bolt to connect a potentially troublesome area of rock to good ground. When in place, the steel plate would fit against the rock. This more modern method replaced the earlier method of adding timbers to support problem areas. The mines were generally noted for their solid ground condition; only a few areas ever needed this type of preventive pinning.

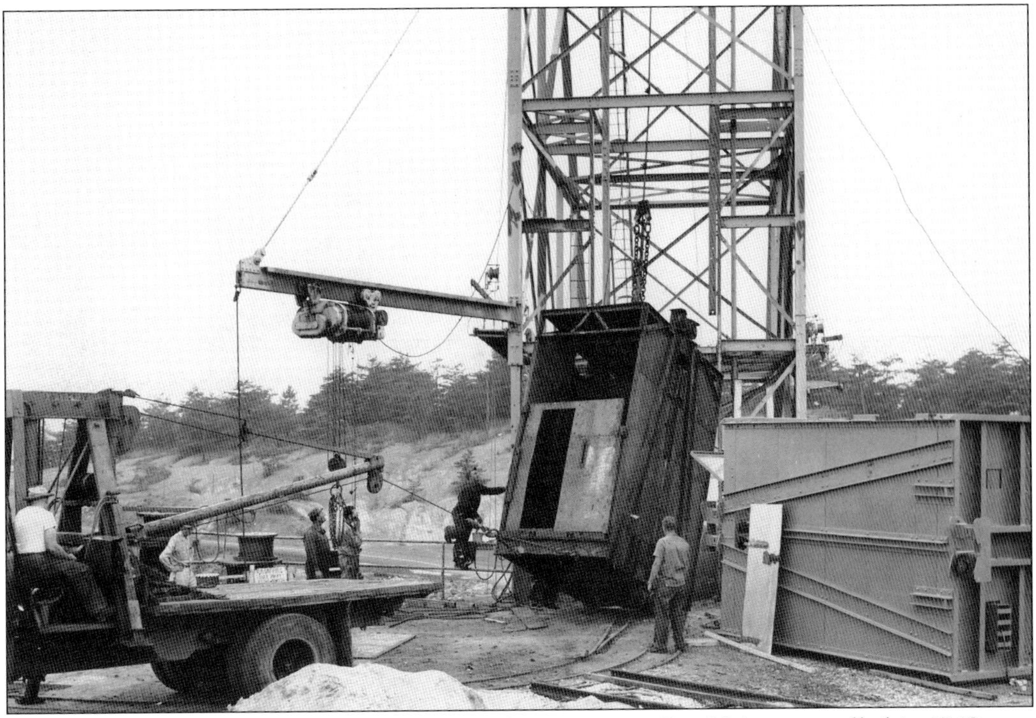

Replacing a Man-Cage, 1962. The original man-cage at Boyd Mine, installed in 1948, was replaced in 1962 by a new cage (right) fabricated in the TCC shops. The new 6-by-10-foot cage weighed 10,360 pounds, was nearly 14 feet high, and could carry as many as 40 men at a time. A door in the floor also allowed for the transportation of large objects.

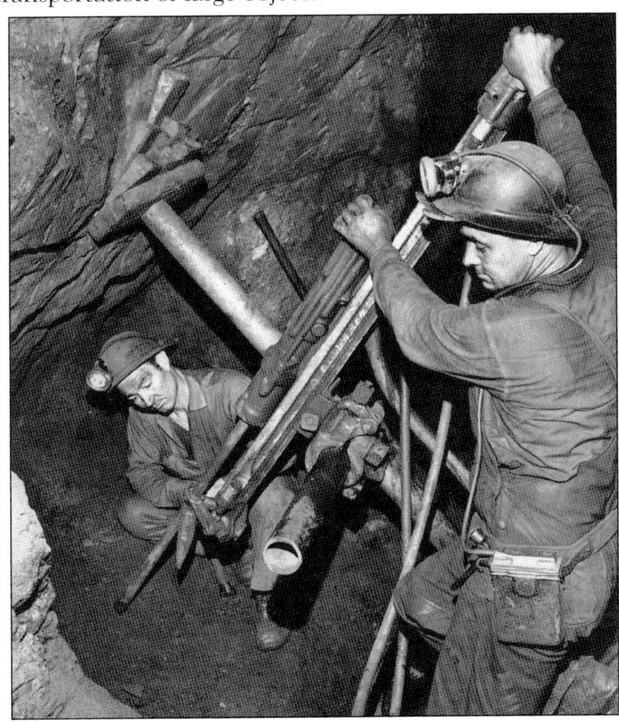

Exploration at Calloway, 1963. At a depth of 2,100 feet, the 24th level of Calloway Mine was already 500 feet below sea level. In 1963, however, TCC began to evaluate the possibility of going deeper. In this photograph, drillers W.C. Patterson (kneeling) and Norman Pittman work 50 feet below the 24th level as part of the team exploring the size and shape of the ore body beneath that level.

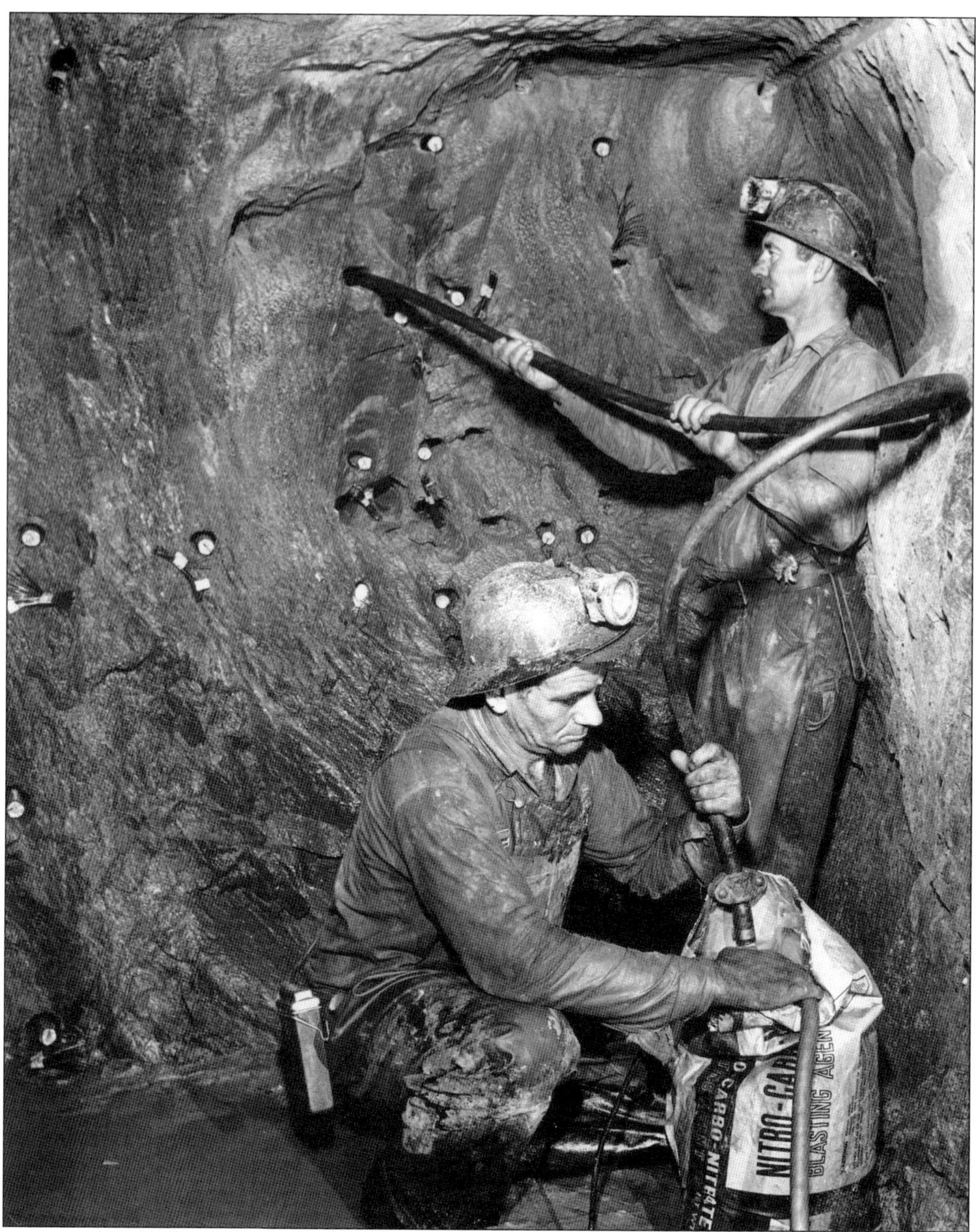

UNDERGROUND BLASTING, 1962. By 1962, dynamite was on its way out as the most-used blasting agent in the TCC mines, and what the miners jokingly referred to as "fertilizer" was on its way in. The new explosive, a granular mixture of ammonium nitrate and fuel oil, was not only cheaper, but also eliminated the severe headaches that could be caused by the nitroglycerin in the dynamite. In this photograph, Roy Graves (left) and Paul Cavender prepare for blasting by blowing granules into deep drilled holes. Single sticks of dynamite placed at the outer ends of the holes acted as primers for blasting, which was done only when all personnel were safely out of the mine.

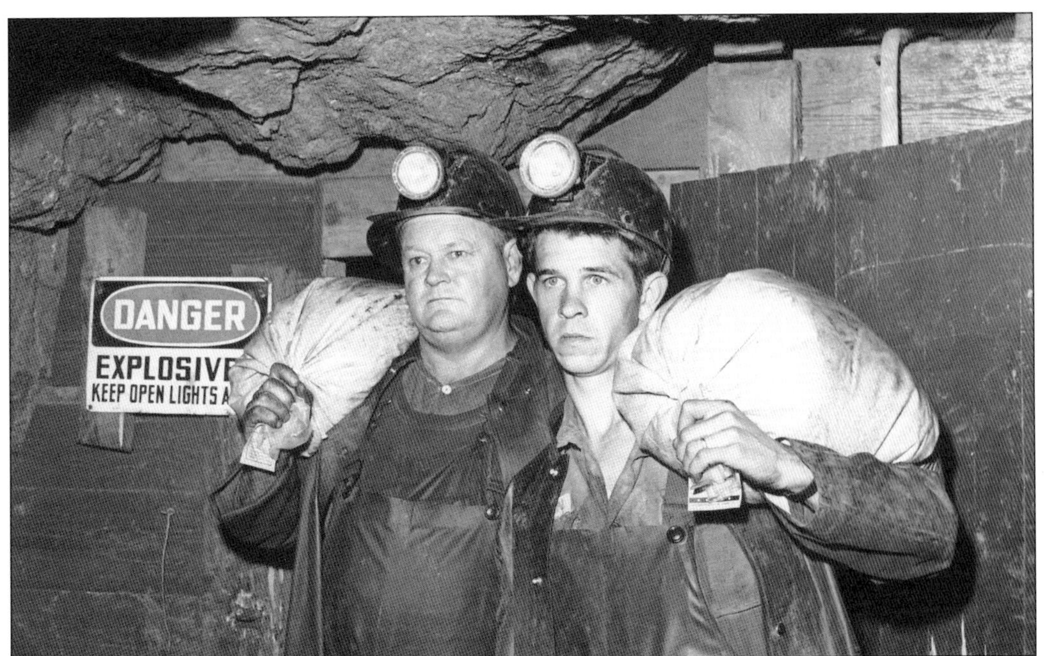

Drilling at Mary Mine, 1964. Above, drillers W.L. Morrow (left) and Russell Amburn carry explosives to begin their work in Mary Mine. Below, evening-shift drillers Elmer Swafford (left) and A.B. Dickey prepare to drill a hole in a drift round. At right is evening stope foreman Willard Chancey. In other areas of the mine, evening-shift crew members were working at the same time to maintain track, repair equipment, and clear rock from previous drilling.

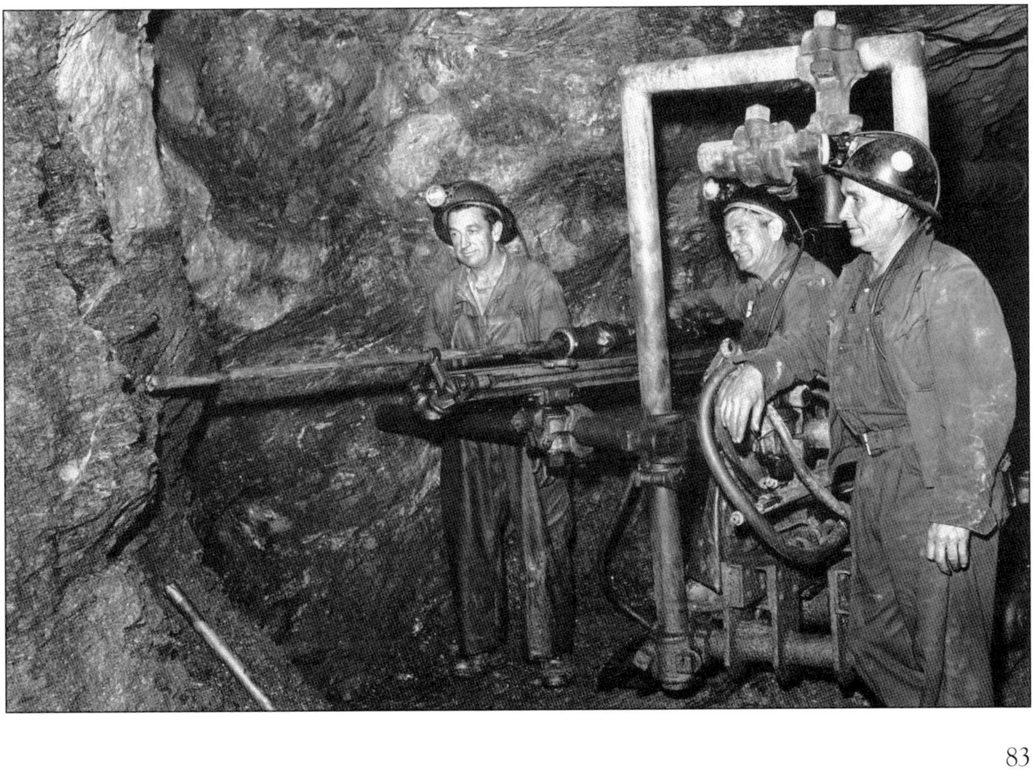

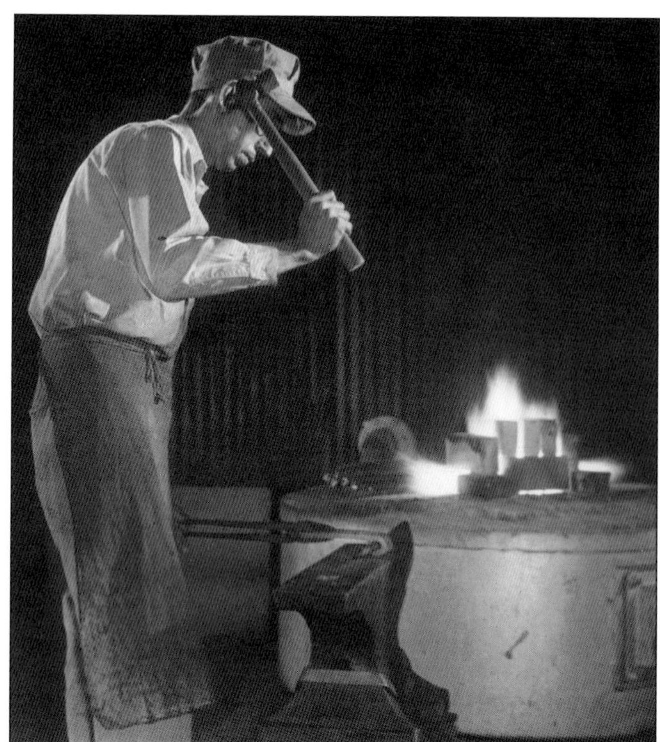

Blacksmith Shop, 1967. James Phillips, at work in the blacksmith shop, was one of five brothers who had followed their father, Nathan Phillips, into employment at TCC. James Phillips first joined the railroad crew in 1940 but spent the majority of his career as a blacksmith, retiring in 1982 after 42 years on the job. His father had compiled a 50-year service record at the time of his retirement in 1955.

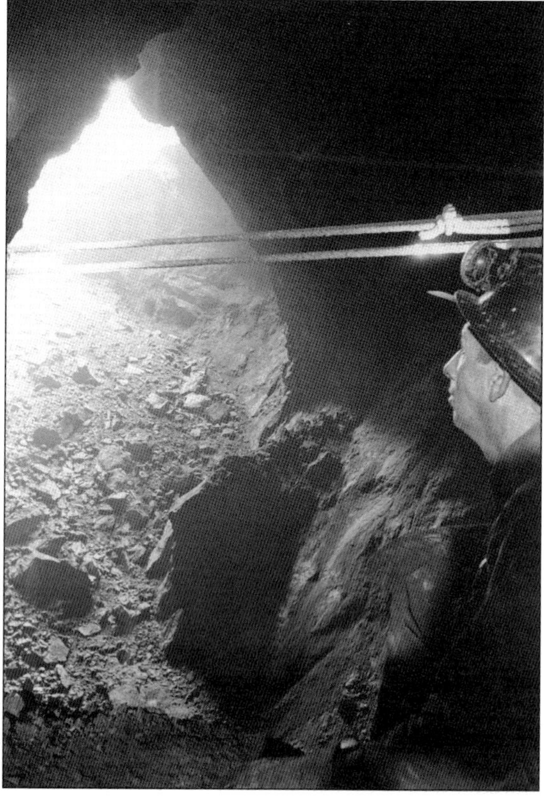

Underground Daylight, 1963. Stopes in the mines, great underground cavities formed when ore was extracted, almost never saw daylight. Due to an unusual small surface-level hole, however, light could penetrate to the third level of Eureka Mine, nearly 500 feet down. In this photograph taken with nothing but natural daylight and a flashgun, stope foreman R.L. Martin stands in an opening in the stope between the first and second levels of the mine.

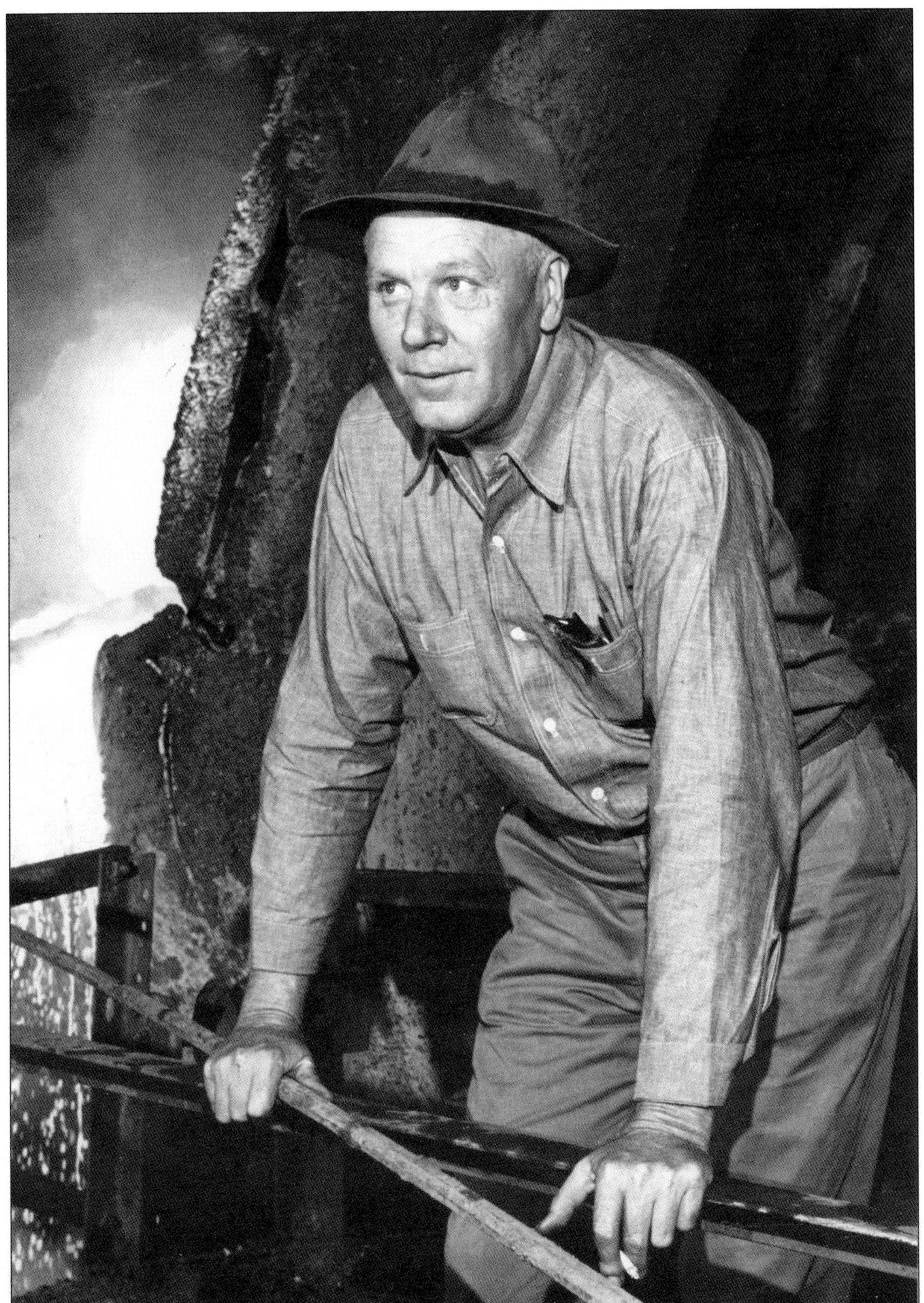

CONVERTERS, 1963. Three months after joining TCC in May 1943, Lester White was transferred to the smelting department, where he had risen to the position of converter foreman when he celebrated his 20th safety anniversary in 1963. His job required a trained eye to see the differences in the colors of molten metal and slag, which was key to determining when the slag had been poured off, leaving the purified copper behind.

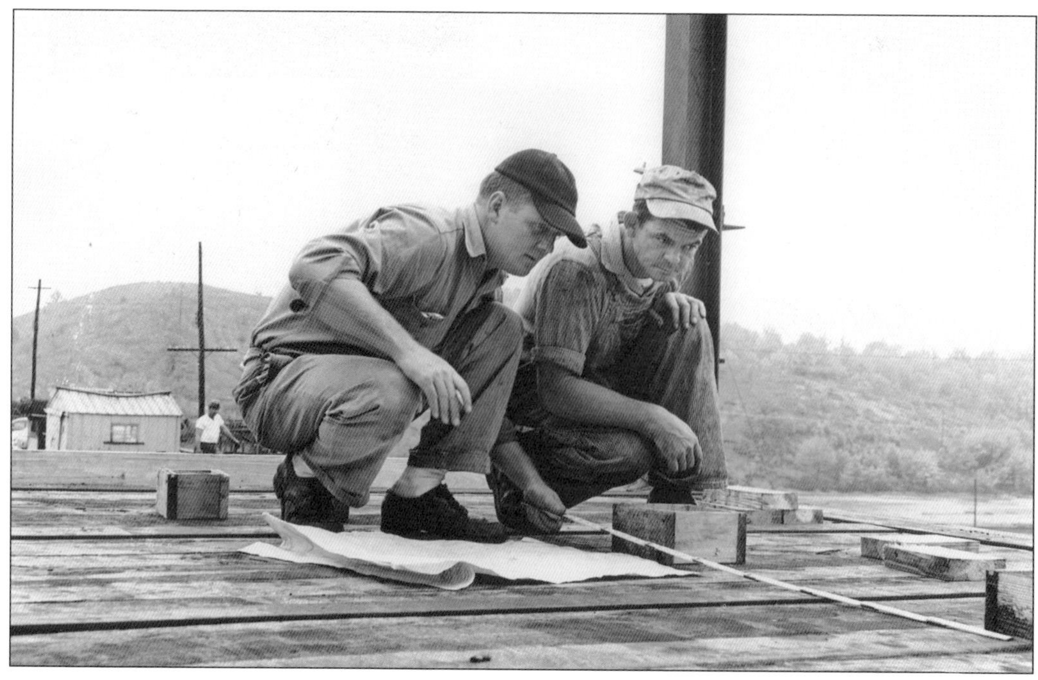

Repairs and Rebuilding, Early 1960s. Above, in September 1960, carpenter foreman Fred Haren (left) and Tommy Campbell check the distance between forms for pouring concrete at the new liquid sulfur dioxide building, which was being rebuilt after the old building was destroyed by fire earlier in the month. Below, in July 1961, carpenter Charles Cole removes a wood structure that had been used to form a new arched roof during a complete rebuilding of the reverberatory furnace. Supervising at right is general carpenter foreman Frank G. Beaver. The TCC copper-making facilities were shut down for three weeks, the longest smelter shutdown in the company's history to that date, while the furnace was rebuilt to adapt to upcoming changes in the smelting process.

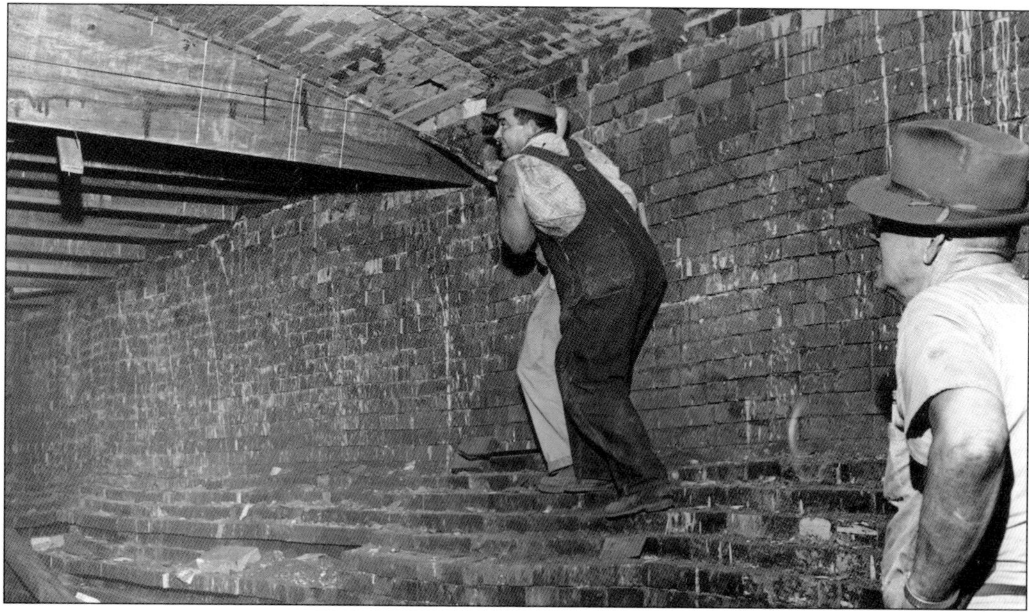

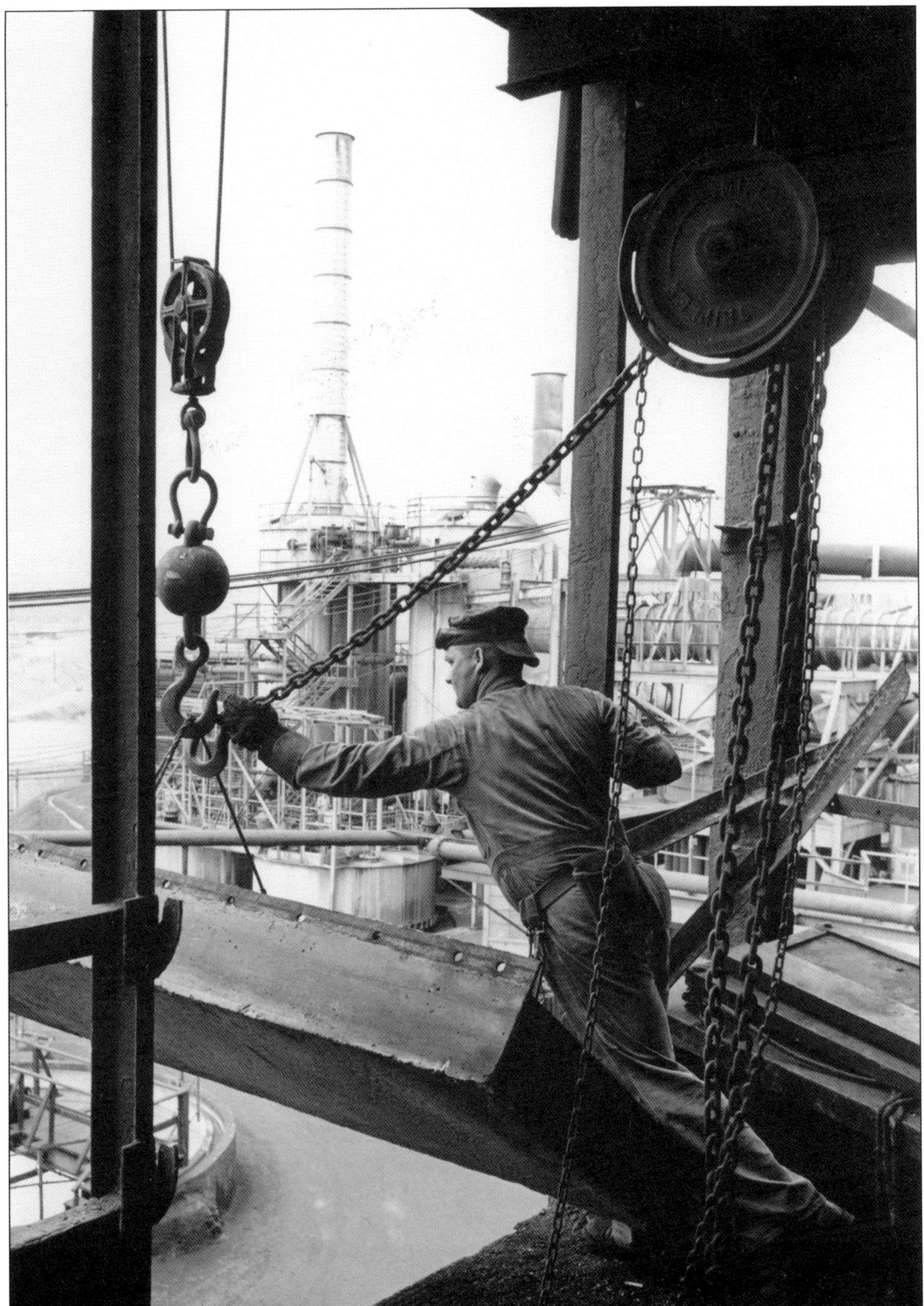

Roaster Repairs, 1960. On May 25, 1960, the Copperhill roasters were shut down temporarily for maintenance, primarily repairs to the flues of the dust-collecting system. The purpose of the dust-collecting system was to remove and recover a large percentage of iron dust from sulfur dioxide gas for use in the production of iron sinter. In this photograph, rigger Charles Kyle leans out to hook onto a pre-cast concrete slab that had been hoisted to the burning floor of the roasters for placement in the roof of a flue. In the background is the No. 3 contact acid plant, completed in 1954.

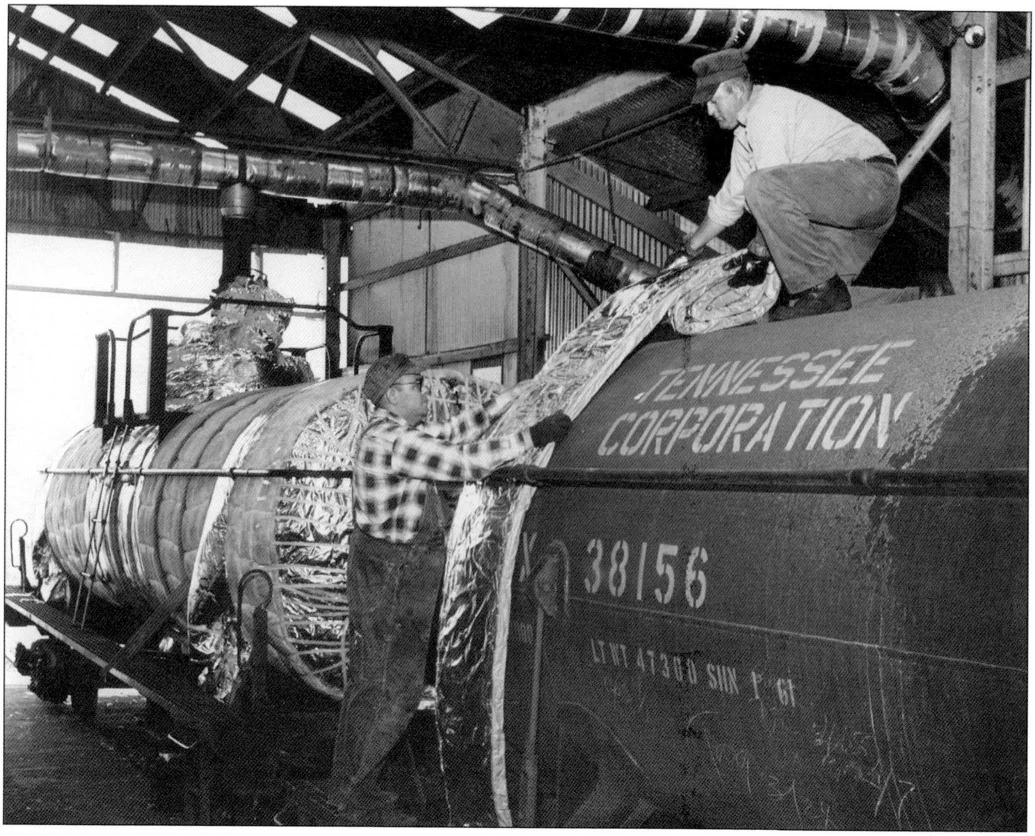

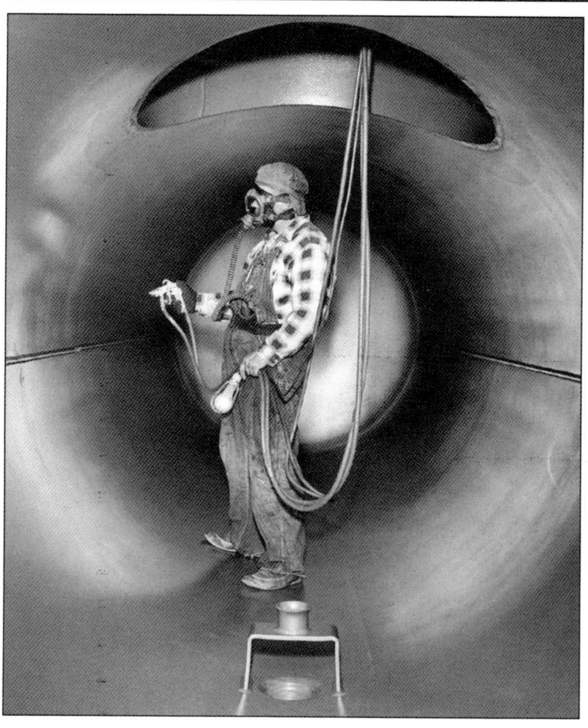

Lining Acid Tank Cars, 1961. Above, railroad department painter Hadley Shields (left) and helper Gene Wade Whittenbarger wrap insulation around an acid tank car in preparation for the application of a heat-hardened, acid-proof coating to the inside of the car. The insulation was needed to help retain high temperatures inside the car after the application of the lining material. Below, inside a tank car, Shields prepares to spray a coat of the brownish-yellow lining material, a synthetic resin that would then be baked at temperatures up to 400 degrees Fahrenheit. The result was a hard surface similar to Bakelite or the coating of household appliances.

London Mill, 1962. Leroy Ledford supervises the filtering process that removed water from iron sulfide pulp at the London Mill flotation plant, creating a concentrate that would be transferred to bins by conveyor belt. By December 1962, when this photograph was taken, London Mill was the largest producer of sulfide concentrates in the United States.

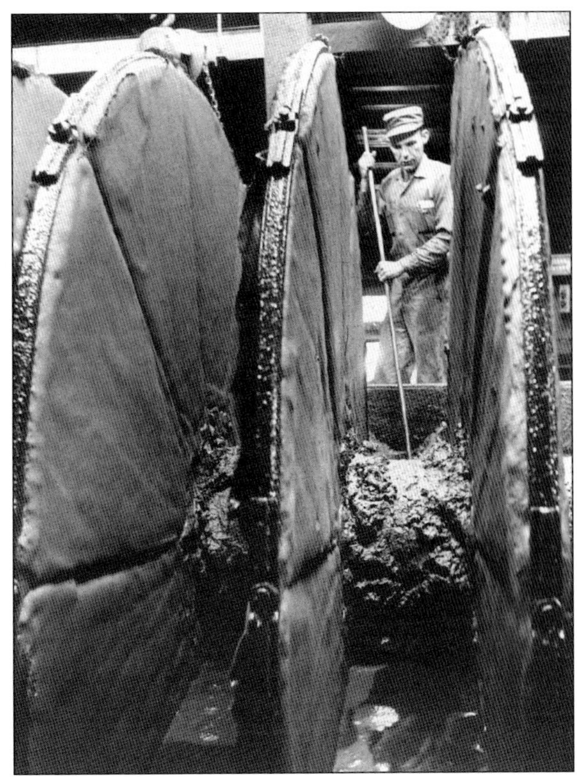

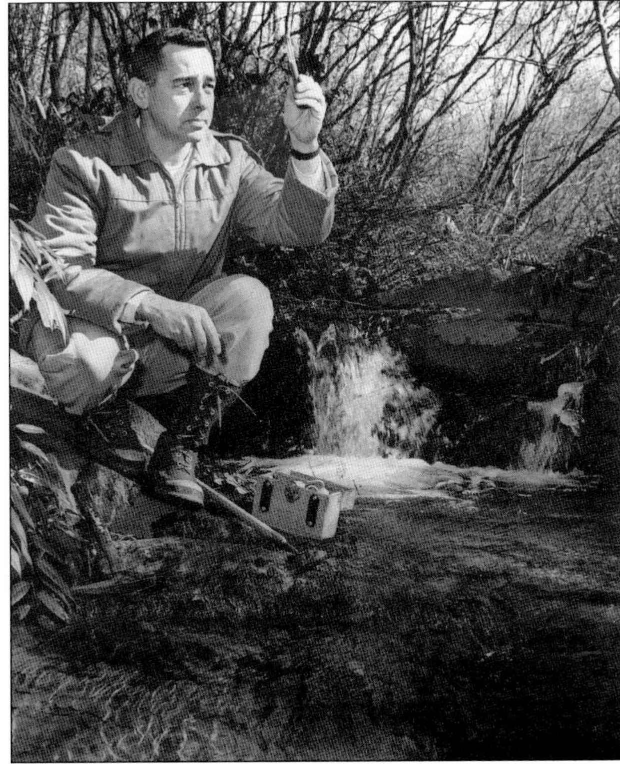

Geochemical Survey, 1965. Chief geologist Maurice Magee holds a sample drawn from a creek bed in a geologically promising area. Samples were taken to the TCC laboratory to be screened for a larger than normal presence of copper and zinc. If a sample proved to be positive, the geologist would move farther up the creek, taking more samples in an attempt to pinpoint the optimal spot for further exploration.

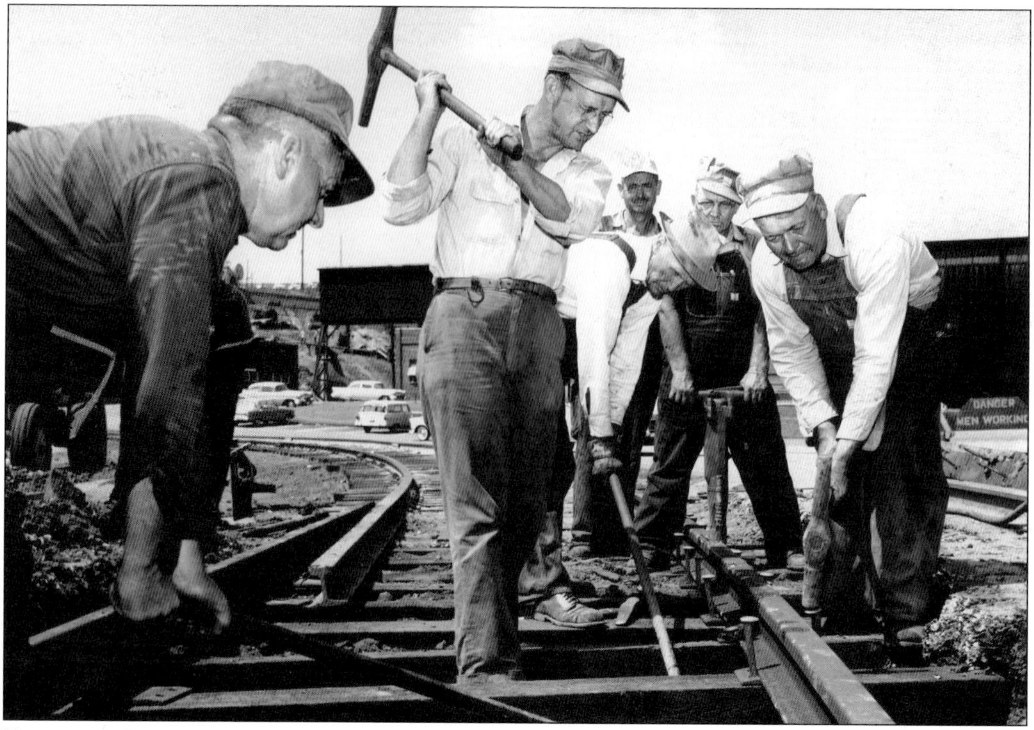

RAILROAD SAFETY RECORD, 1962. In 1962, as the railroad department celebrated a five-year safety record, the track crew celebrated an even more enviable record: 18 years without a disabling accident. Driving spikes in this photograph, from left to right, are Fred Mashburn, Hoyt Sisson, S.C. Crawford, John Swanson, Oliver Duvall, and Fred Chancey.

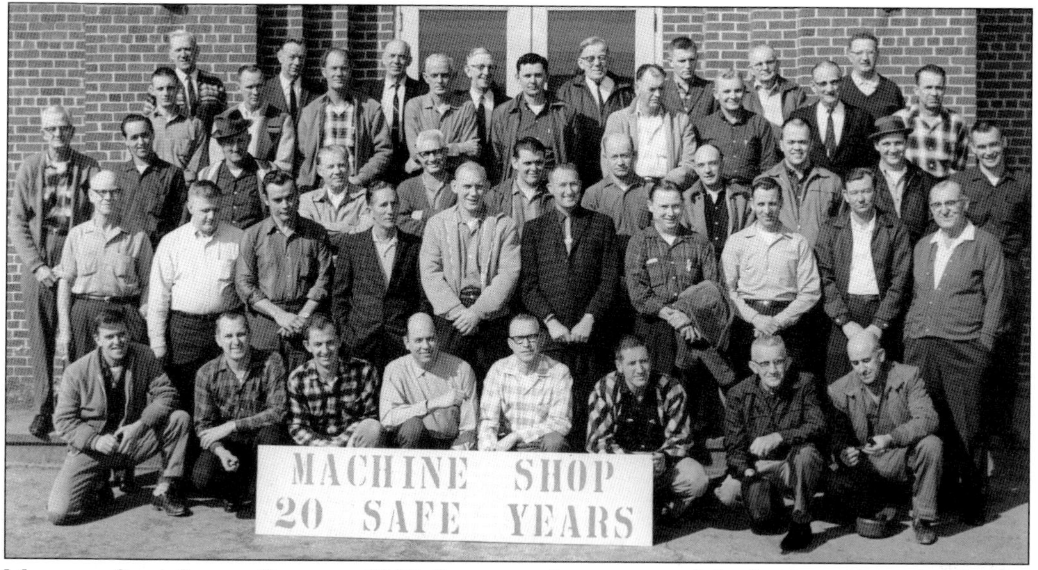

MACHINE SHOP SAFETY RECORD, 1965. In a special meeting on March 10, 1965, workers at the Copperhill machine shop gathered to celebrate a 20-year safety record. The department had not had a disabling accident since March 5, 1945. A number of company executives, in the back row, were present to honor the significance of the occasion.

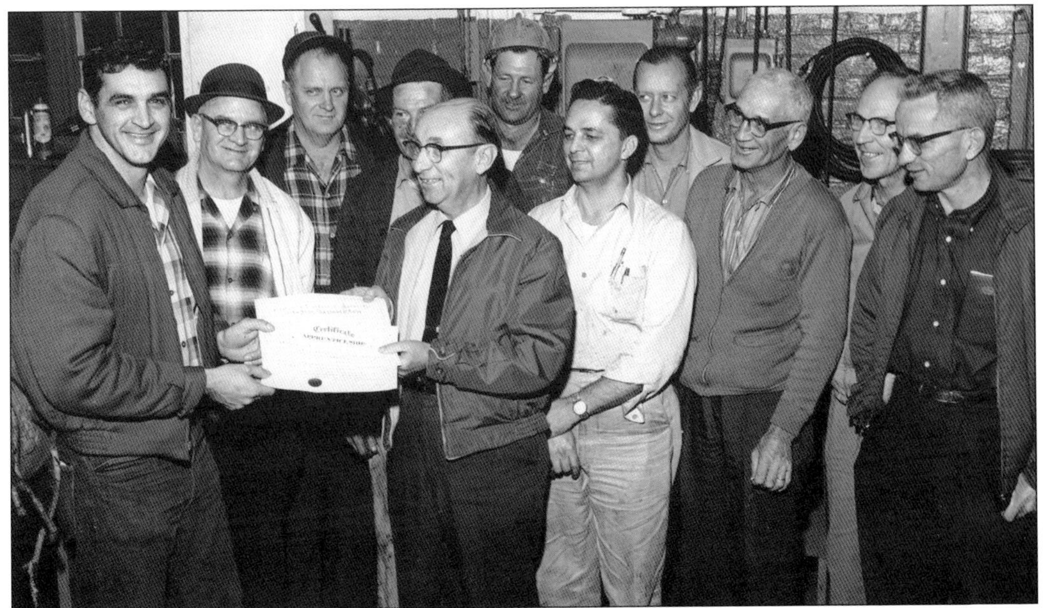

INTERNATIONAL CORRESPONDENCE SCHOOLS GRADUATES, 1960s. Throughout the 1960s, numerous TCC employees continued to take advantage of the company's co-sponsored International Correspondence Schools (ICS) program to improve their job skills and status. Above, in 1966, newly promoted mine electrician Bill Tallent (left) accepts his diploma from mine superintendent Roy Clay. To Clay's right is one of the earliest ICS graduates, mine electrician Bazelle Nelson (see page 77). Below, in 1969, machinist Doyce Loudermilk, who had worked at the Smelter Store before transferring to the mechanical department in 1964, accepts his diploma from chief engineer Forrest Petty. At center is machinist Cyril Queen.

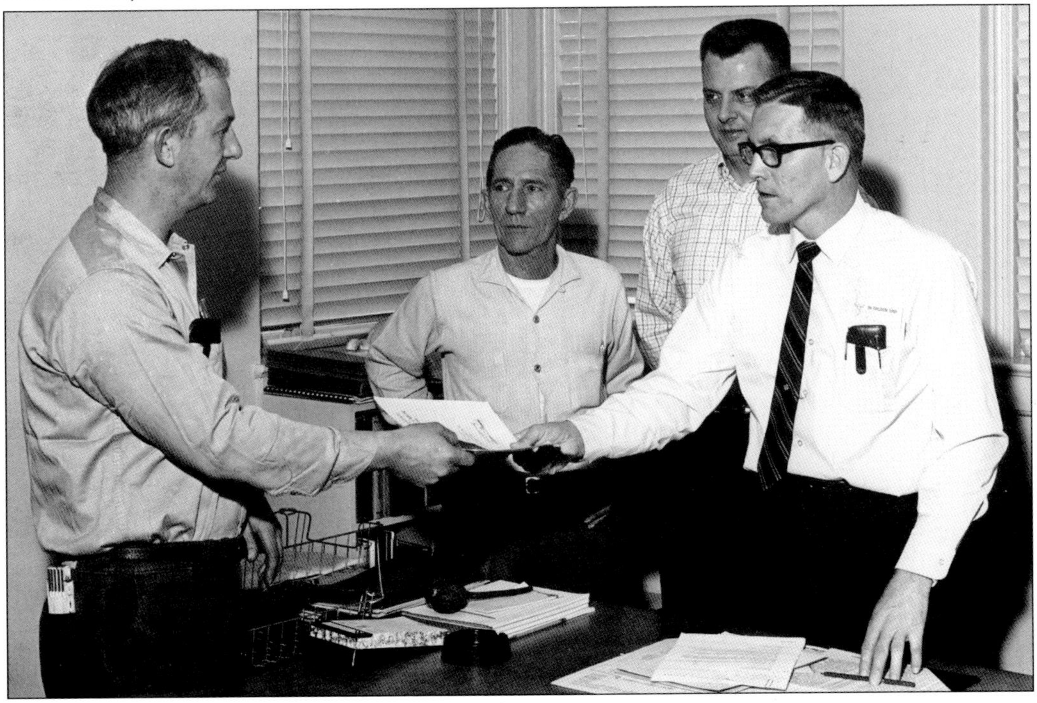

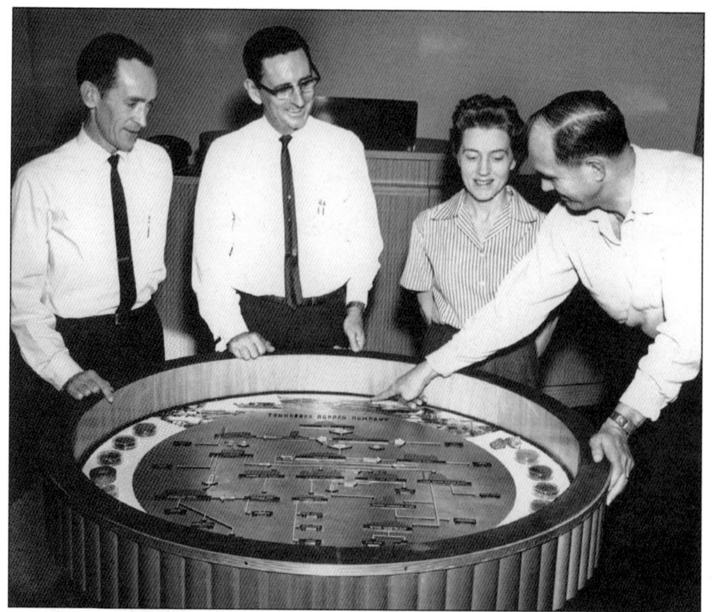

New Lobby Display, 1963. Fred Epperson (right) puts the finishing touches on a new display at the TCC General Office. In the center is a three-dimensional chart of manufacturing facilities at Copperhill. Around the circumference are samples of various TCC products. Looking on, from left to right, are W.J. Taylor, office manager Wallace Whitfield, and Eva Lou Boling. The display is now on exhibit at the Ducktown Basin Museum.

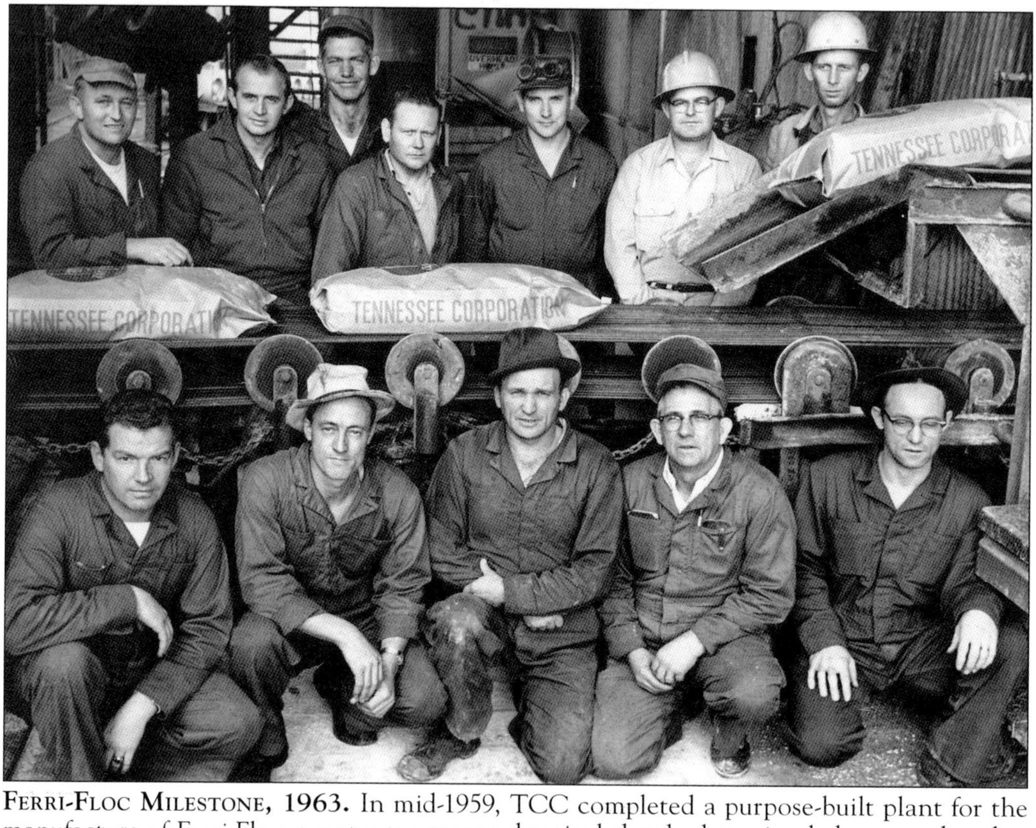

Ferri-Floc Milestone, 1963. In mid-1959, TCC completed a purpose-built plant for the manufacture of Ferri-Floc, a water treatment chemical that had previously been produced at other Tennessee Corporation plants using raw materials from TCC. In February 1963, the company celebrated the shipment of its one-millionth 100-pound bag. Ferri-Floc, in demand by large municipal water systems, was also shipped in bulk form and in drums.

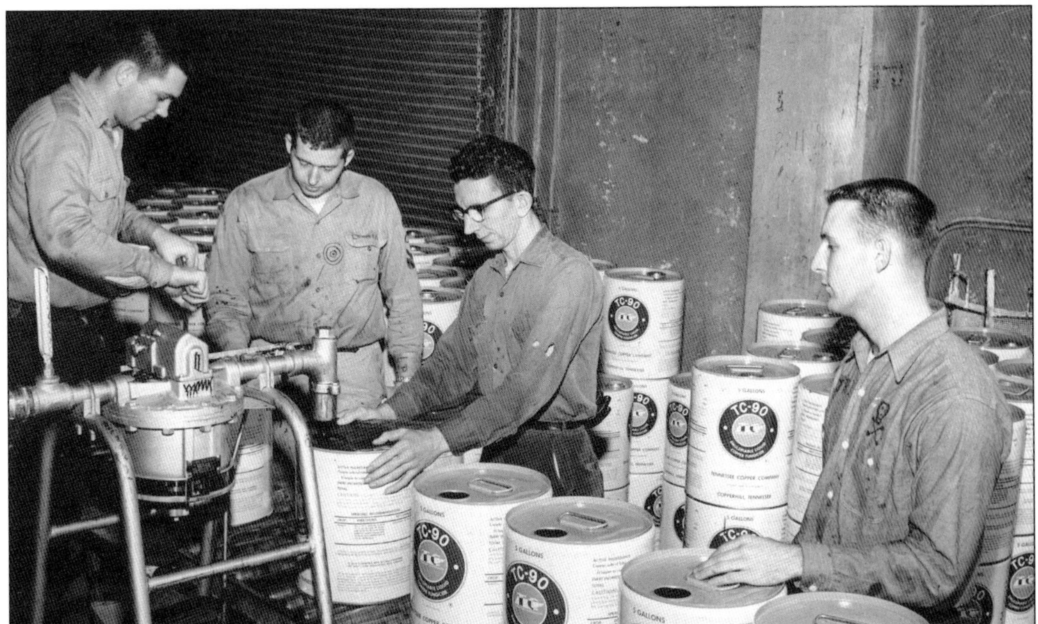

New Copper Fungicide, 1963. Workers at the organic chemicals plant prepare cans of Tennessee Corporation's newly trademarked liquid copper fungicide, TC-90, for trial marketing in the 1963 season. From left to right are William J. Towery, Leonard Parris, Eugene Phillips (at the automatic filling machine), and Donnie Dockery.

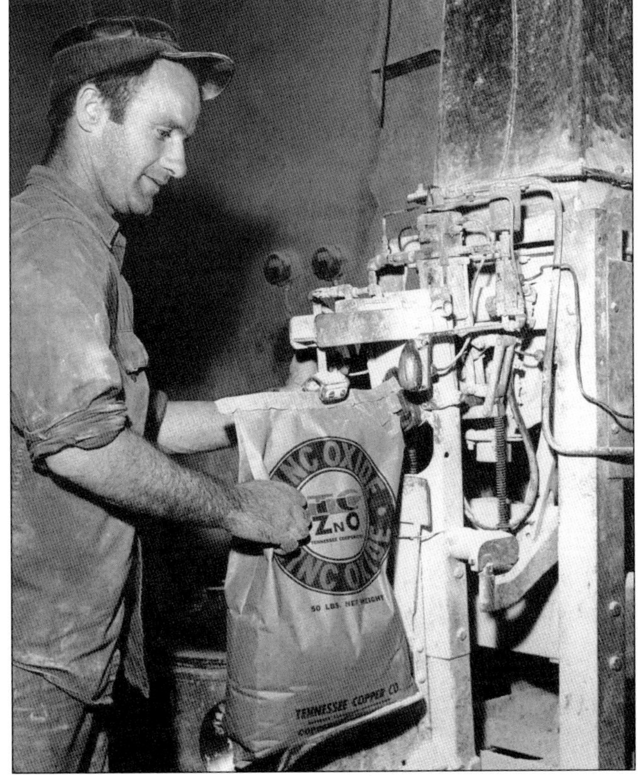

Bagging Zinc Oxide, 1966. At the sodium hydrosulfite plant, bagger James Farmer fills a 50-pound bag with finely powdered zinc oxide, one of the finished products of the plant's multi-step manufacturing processes. Although some of TCC's zinc oxide was marketed for agricultural purposes, the vast majority was sold to rayon manufacturers.

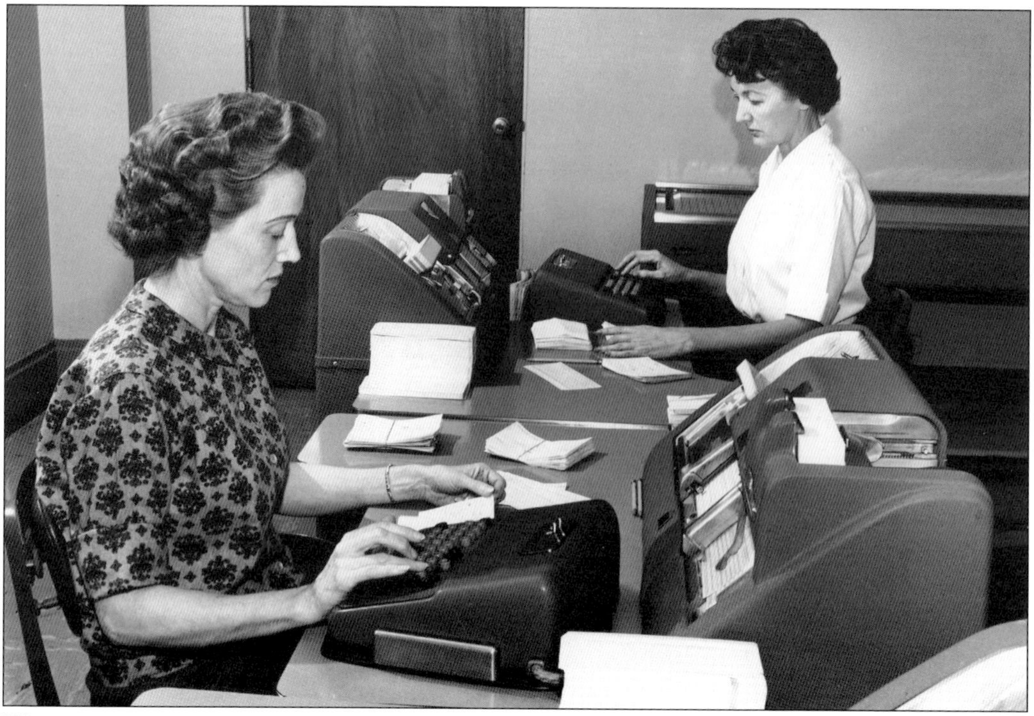

Women at Work, 1960s. Due to the nature of the industry, as well as the cultural constraints of the times, few women were ever employed at TCC. Most held clerical positions, like longtime payroll clerks Eva Lou Boling (left) and Ann Mull, above. Others worked at the Smelter Store or in the laboratory. Below, shipping department clerk Gladys Arquette checks freight bills in preparation for the shipment of sulfuric acid.

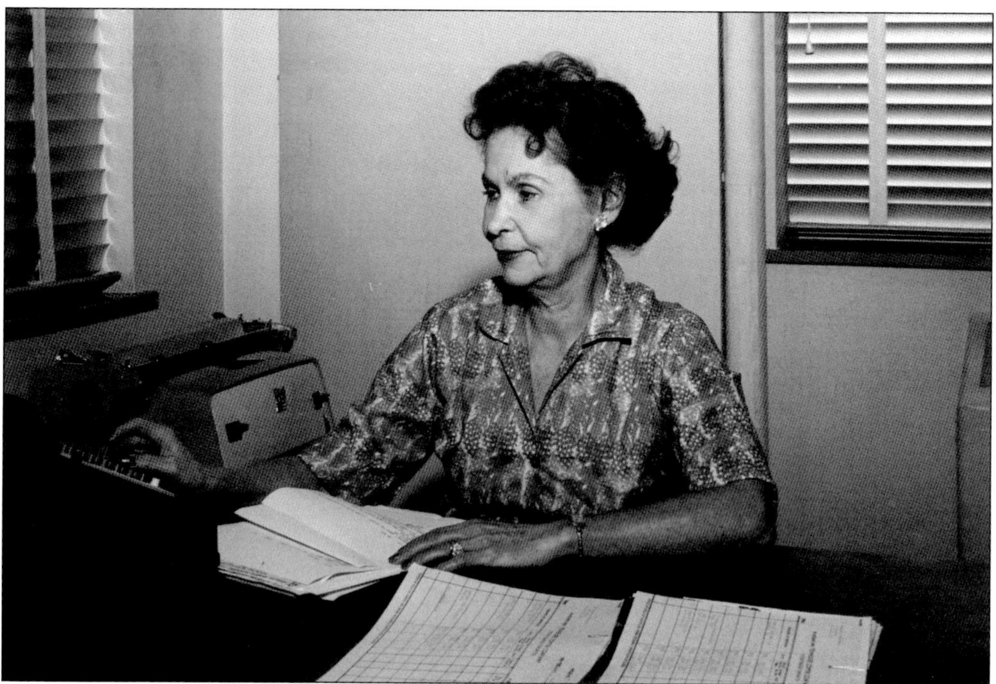

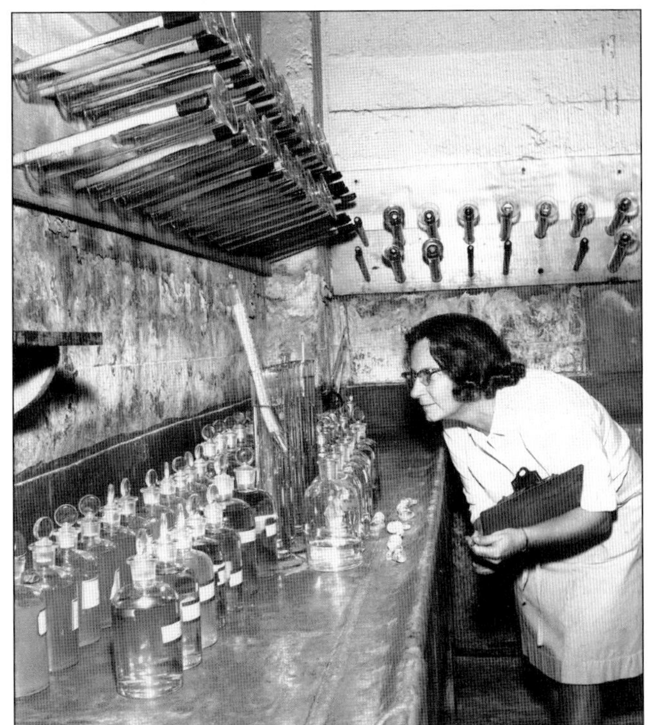

Chemistry Laboratory, 1964. Several women worked in the TCC laboratory at Copperhill, where samples of all the company's products were examined before shipment. Among the responsibilities of laboratory chemist Loleta Adams was to measure sulfuric acid samples for the correct acidity levels. One sample was tested, then retested as a check on the first test, from every sulfuric acid tank car leaving the plant.

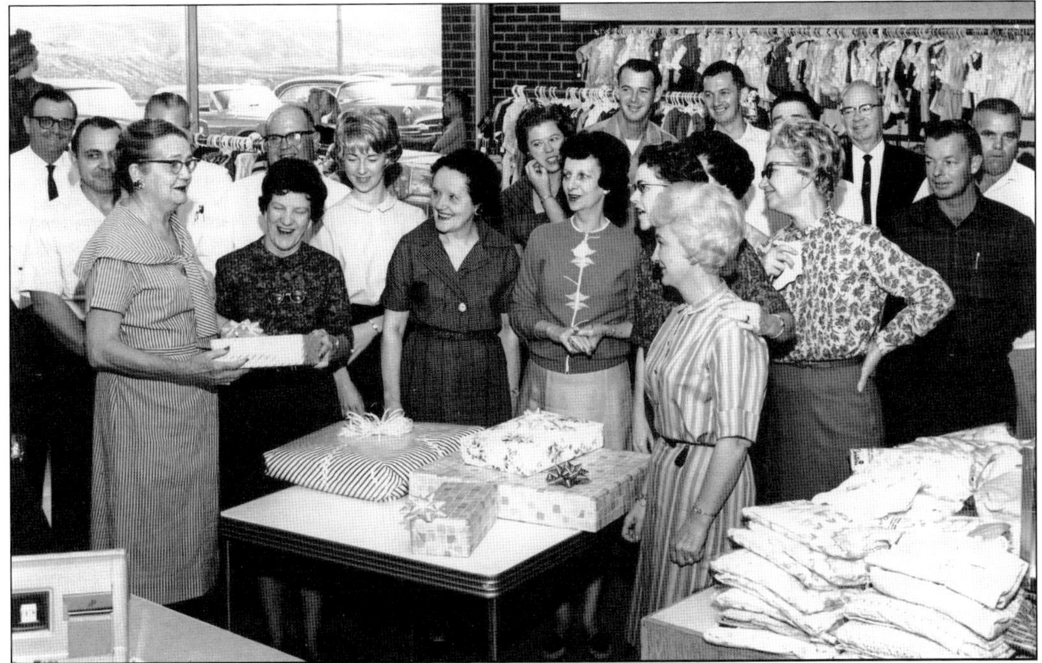

Smelter Store Retirement, 1963. Surrounded by Smelter Store management and staff, Anna Belle Cunningham (left) accepts a retirement gift from co-worker Susie Barrett. Cunningham, a familiar face at the Copperhill store for many years, had originally gone to work for the TCC Ducktown store as a teenager in the early 1900s. Her brothers Earl and Cliff were TCC lead burners who also spent long careers with the company.

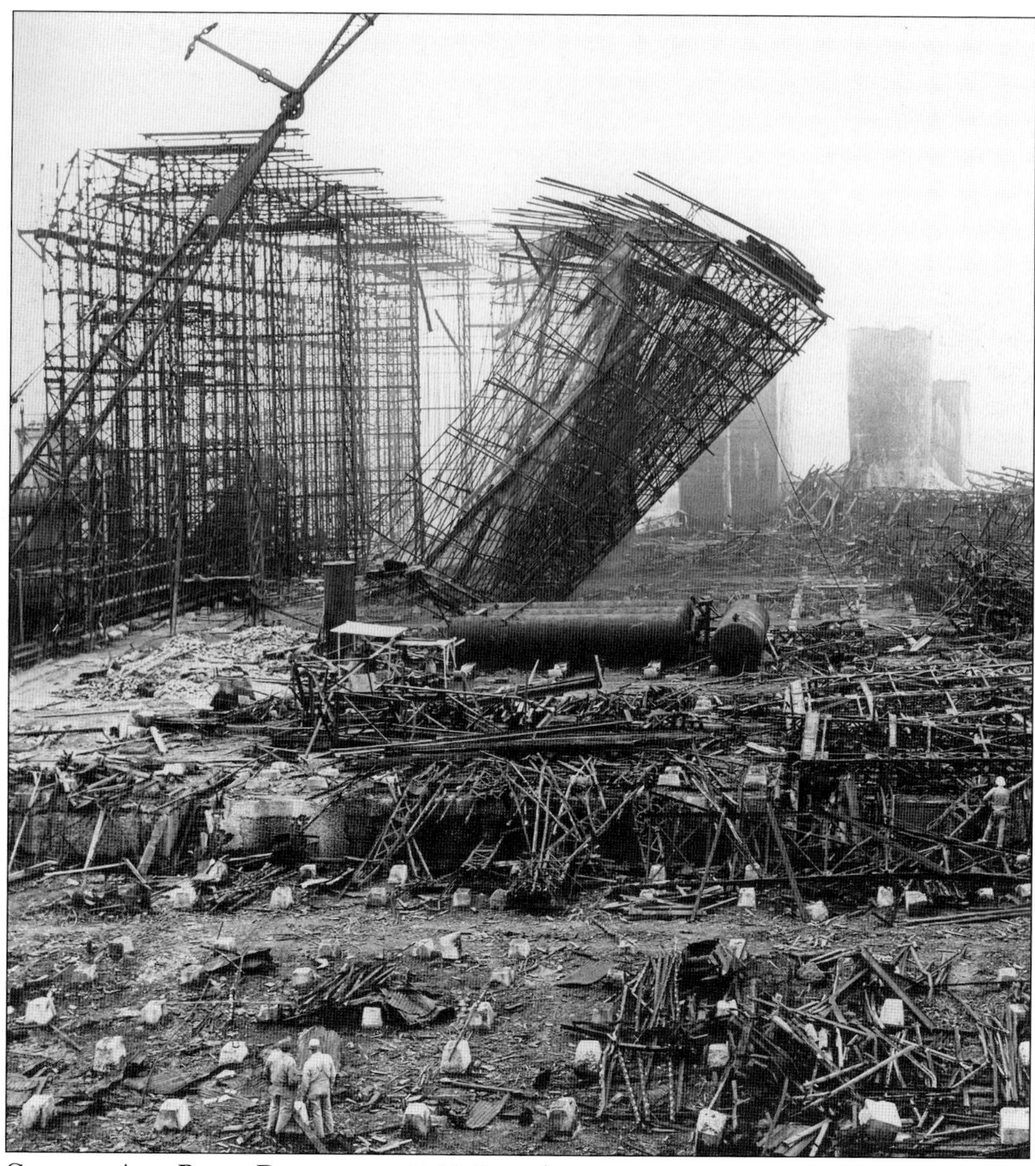

Chamber Acid Plant Demolition, 1965. By mid-1964, TCC's acid department had converted entirely to the contact acid process, and the original lead chamber plant that had dominated the Copperhill skyline since 1907 was marked for demolition. Since the chamber plant was surrounded by buildings that were still in use, detailed planning was required for the rigger crews to take down the giant structure without damaging the surrounding ones. Some steel sections had to be removed by crane to make room for the rows of 70-foot columns to fall when they were finally severed from the metal base, as seen here. It was also necessary to take into account that, due to the architecture of the structure, different portions would fall in different directions. The dismantling of the 4.5-acre plant began in July 1964 and was completed by April 1965, although final salvage and cleanup were not completed until early 1966. This photograph is thought to have been taken in January 1965.

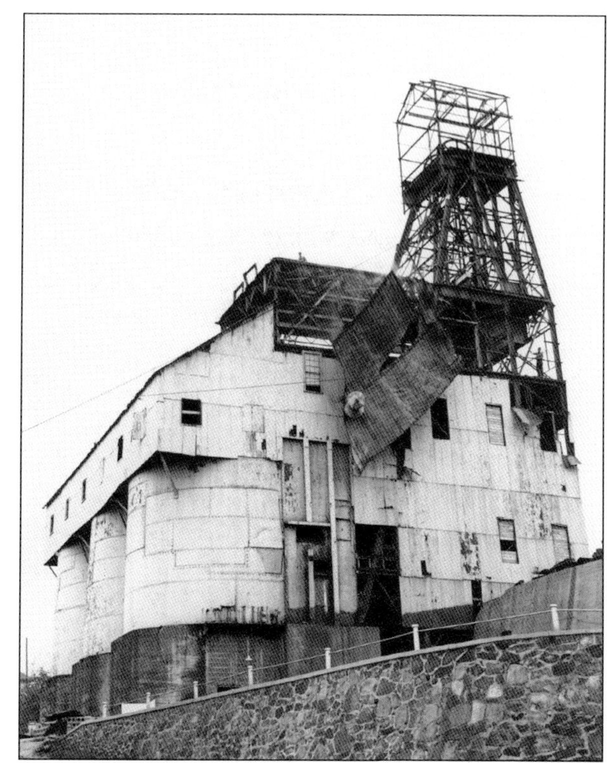

DEMOLITION AT BURRA BURRA, 1966. On August 23, 1966, the first sections of metal siding (right) were removed from the Burra Burra Mine headframe, which had stood on the hill overlooking Ducktown since 1917. By September 8 (below), riggers were using a crane to disassemble the steel skeleton. Although the mine was closed in 1958, the site had remained the headquarters of the TCC mining department, and most of the original structures were still in use. The headframe, however, served no remaining purpose, and the company ultimately decided that it would be too expensive to maintain. Portions of the concrete ore bins were subsequently converted for use as a paint shop.

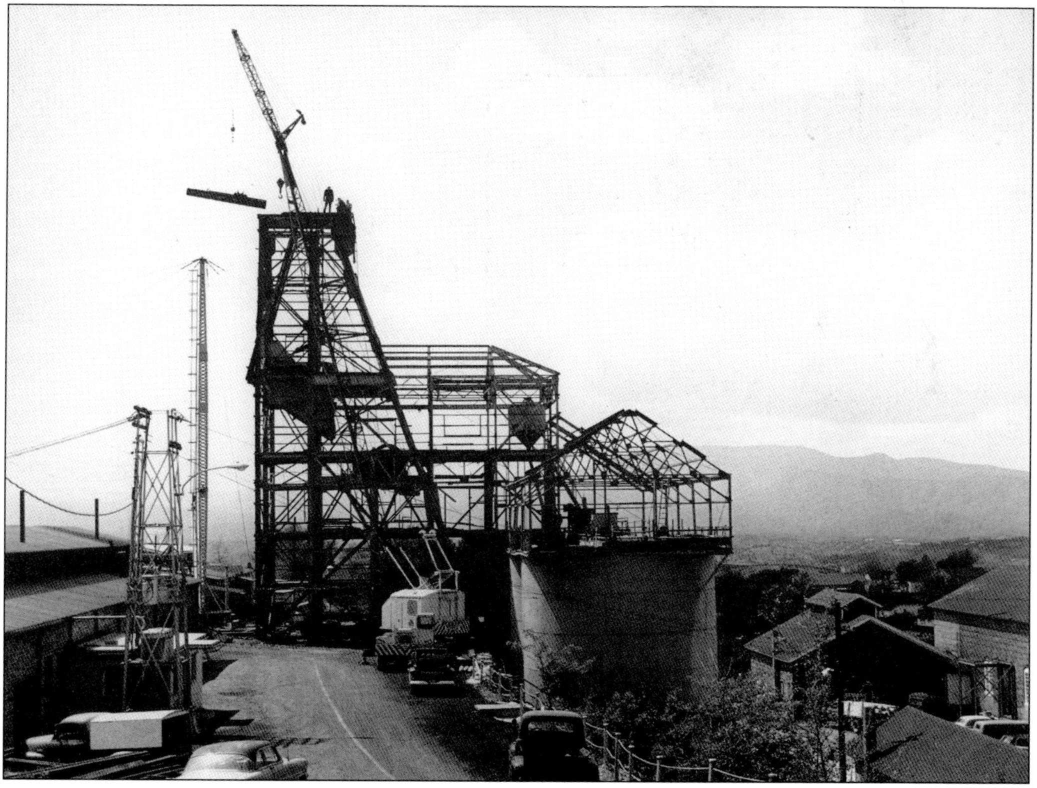

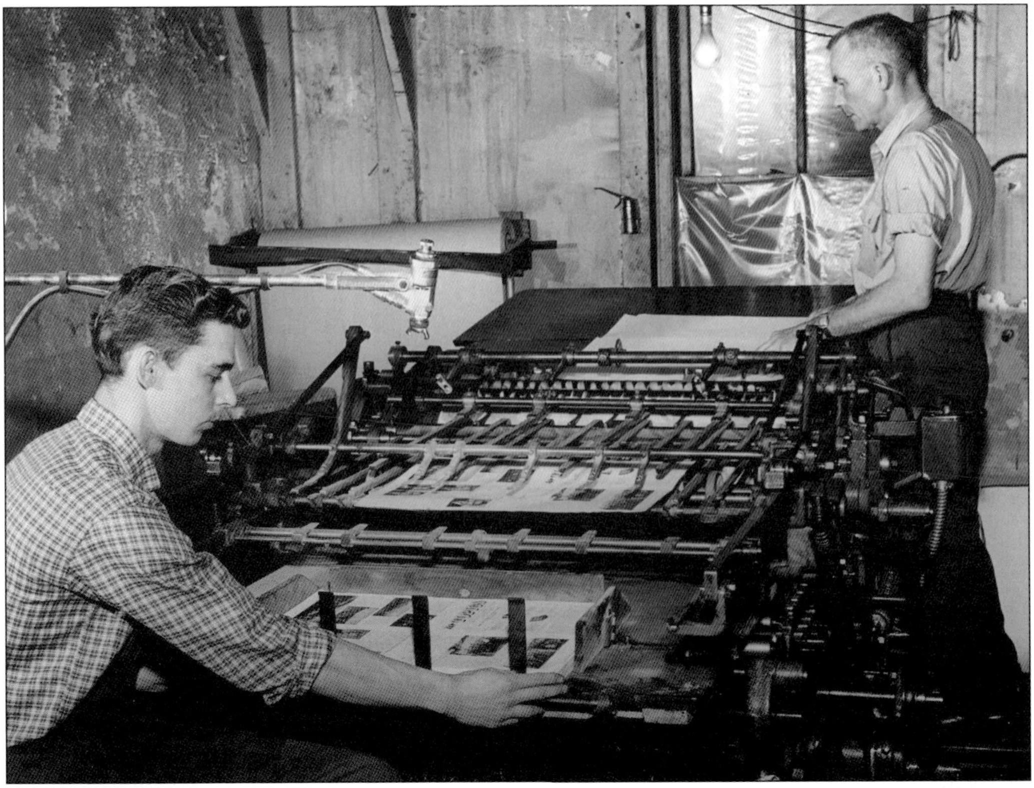

TC Topics, 1962. In 1952, a short-lived TCC newsletter called *The Acid Test* was replaced by *TC Topics*, which was published under that name until December 1969, continuing as *Topics* under later management. Above, Harold Ledford (left) and Copperhill newspaper editor Edward A. Middleton feed pages of the newsletter through the presses of the *Copper City Advance*. Below, from left to right, Roy Kincaid, postmaster J.S. Akin, and Louis Akin prepare copies for mailing at the Copperhill post office.

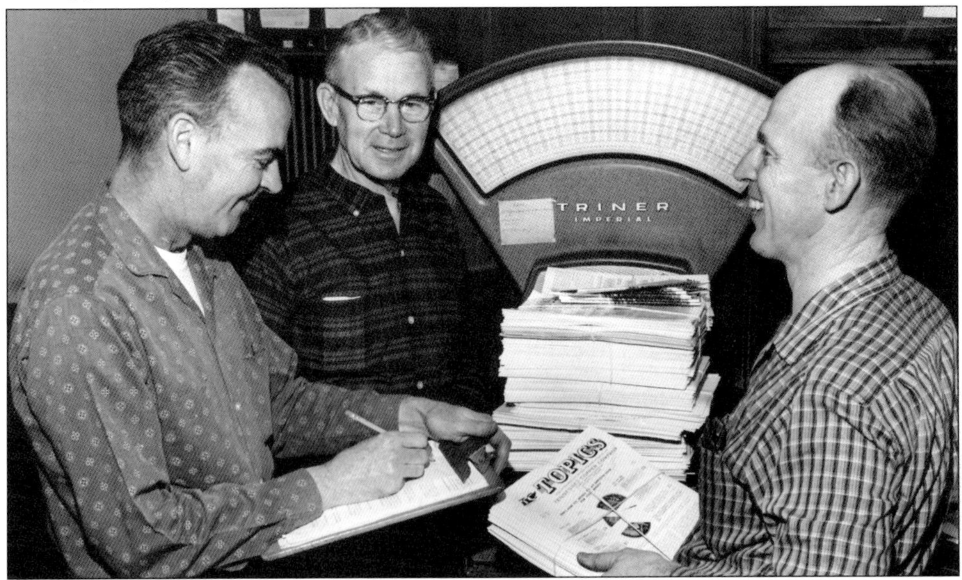

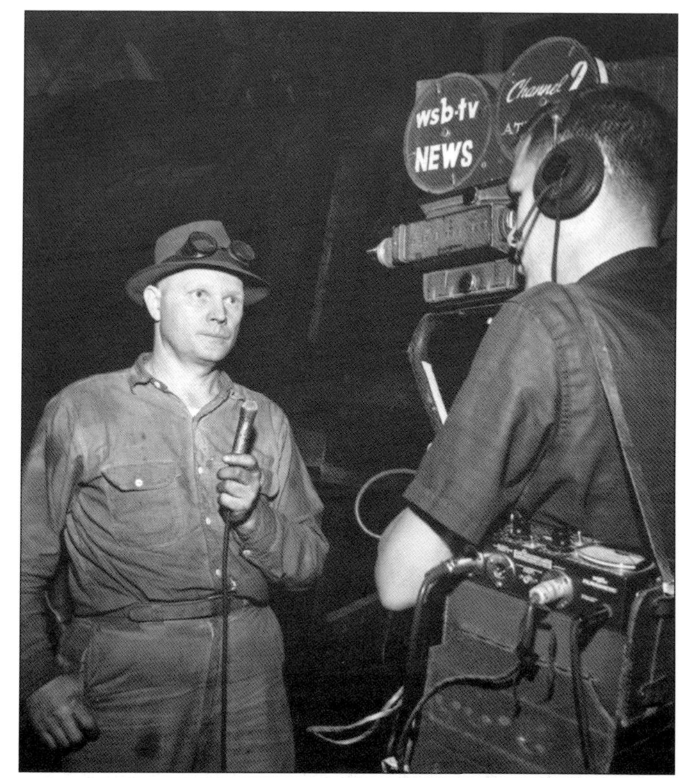

Television Coverage, 1963. During the revival of a longtime boundary dispute in which Georgia attempted to reclaim 60 square miles of Tennessee, Atlanta's WSB-TV came to Copperhill to shoot a segment of a local program about the boundary question. Roughly 10 minutes of the program were devoted to TCC operations, which were almost directly on the current state line within the disputed area. At right, cameraman Dave Riggs interviews smelter worker Woodrow Green. Below, Riggs shoots footage of drillers Oine Towl (left) and Roy Morrow underground at Calloway Mine. Georgia's attempt failed, and all the disputed property remained part of Tennessee.

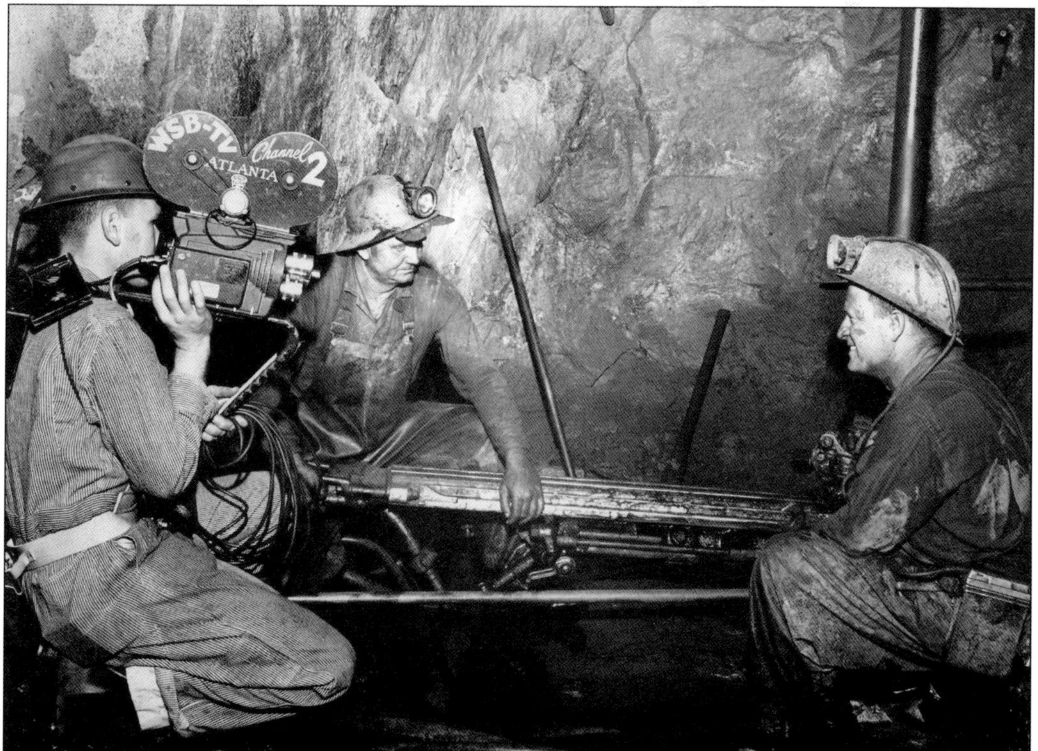

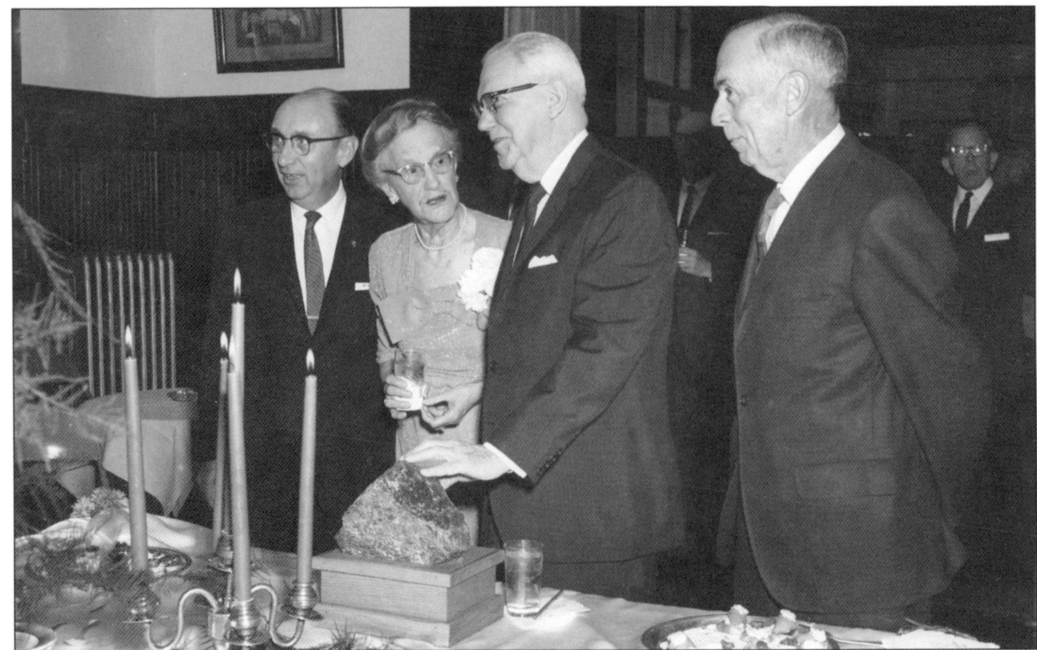

T.A. Mitchell Retires, 1966. To many longtime employees, congenial TCC president and general manager T.A. Mitchell was known simply as "Mitch." It was also the name used by retired Tennessee Corporation president E.H. Westlake in a congratulatory message read at Mitchell's retirement banquet on December 12, 1966. Above, Mitchell and his wife, Lavinia, are flanked by mine superintendent Roy Clay (left) and general superintendent Lamar Weaver. Below, the Mitchells pose with some of the many other executives and wives in attendance. From left to right are (first row, seated) Robert E. Barclay, MaryLu Barclay, Lavinia Mitchell, T.A. Mitchell, Josephine Henegar, and H.B. Henegar; (second row, standing) Pauline Clay, Roy Clay, Dickey Bretz, John Bretz, Bess Hill, H.H. "Happy" Hill, Ava Jones, E.M. Jones, Louise Hawk, Oliver "Brud" Hawk, Barbara Keffer, and W.F. Keffer.

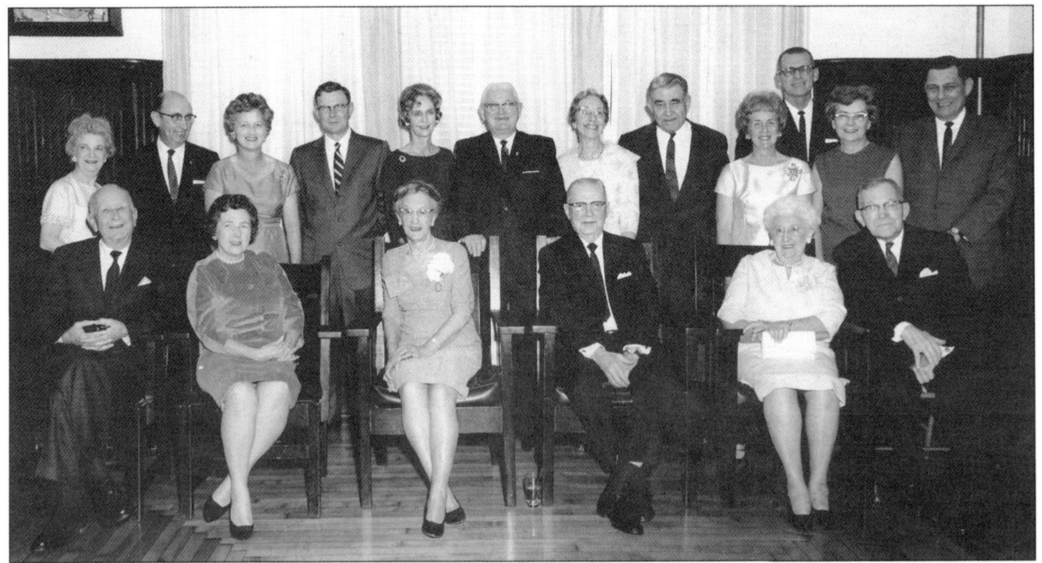

Four

Life after Work

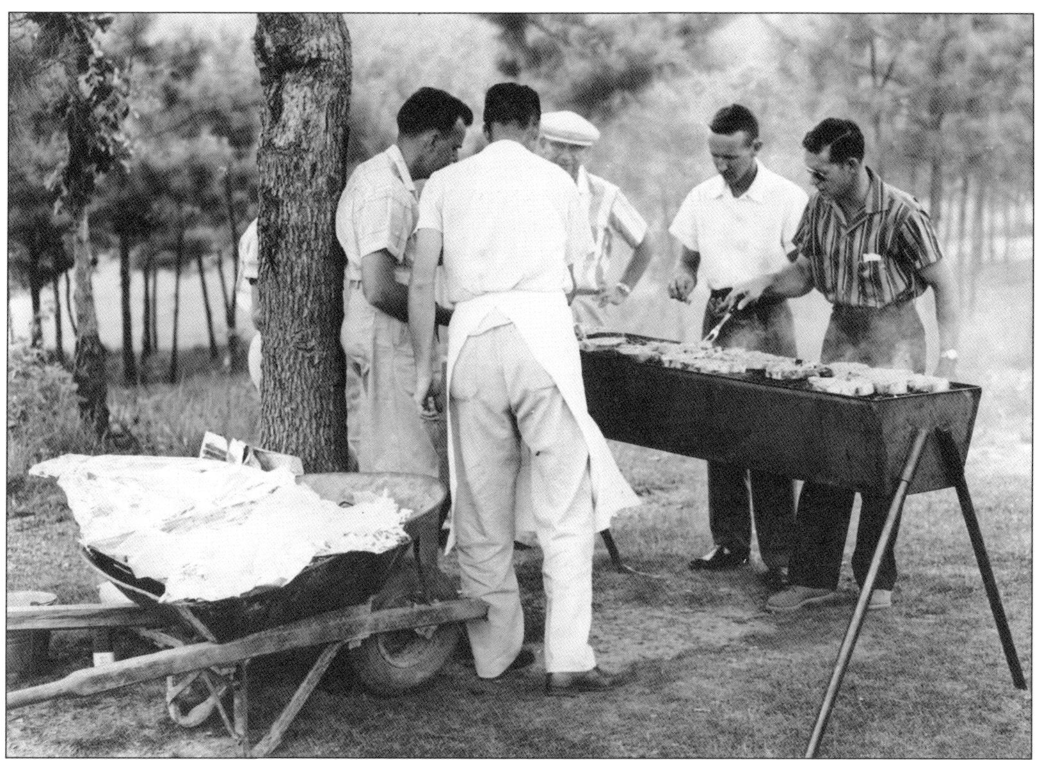

Engineering Department Picnic, 1957. The steaks are on the grill for the July 1957 engineering department picnic, with Cowanee Club manager Vic Hensley (back to the camera) in charge. The others, from left to right, are Guy Livingston, W.C. "Willie" Posey, E.W. Harper Jr., and Robert Lohr, who had designed the charcoal grill. The foil-covered wheelbarrow at left was used to transport the steaks to the grill.

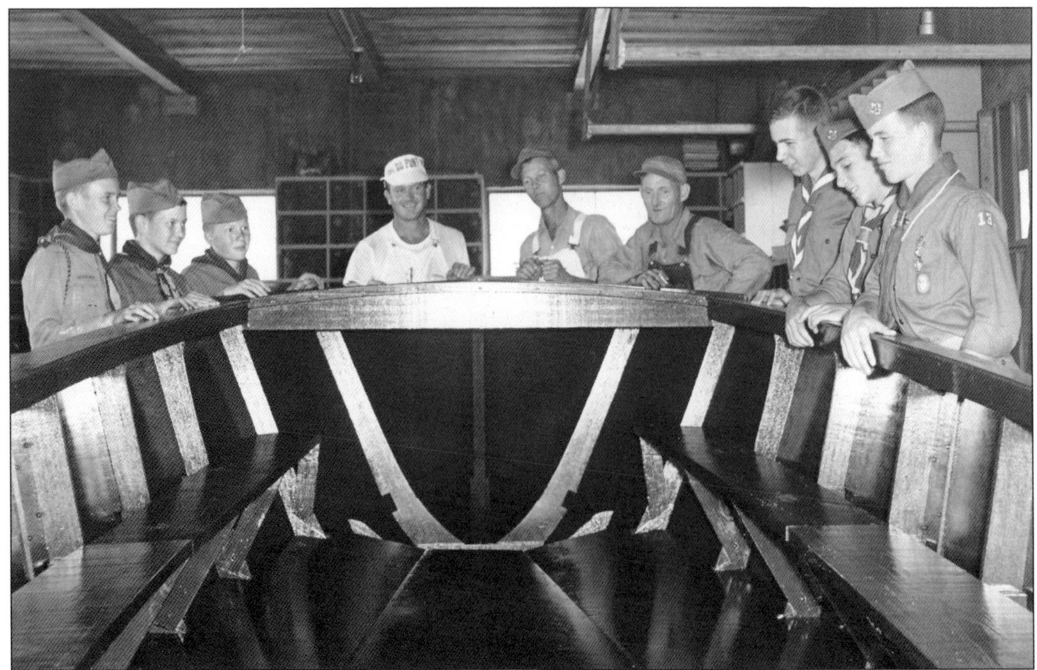

Boy Scout Boat, 1954. Boy Scouts from local troops inspect their just-completed boat in the TCC carpenter shop. Materials for the boat were bought by the Cherokee Area Council, and TCC donated the labor. From left to right are scouts Roger Waters, Billy Henry, and Jimmy Louis Panter; carpenters Fred Davis, Ralph Harper, and Glenn Martin; and scouts Britt Burns, Jon Earl Bell, and George Cobb Jones.

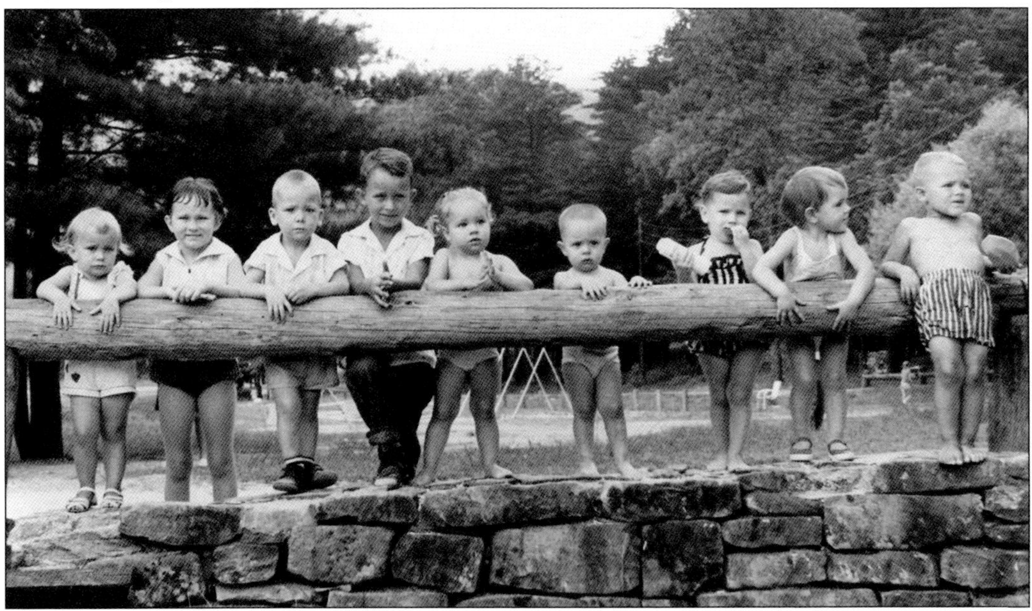

Smelter Store Picnic, 1957. Children of employees assemble for a photograph during the annual Smelter Store outing at Vogel State Park near Blairsville, Georgia. From left to right are Rita Faye Pelfrey, Glenda Kay Bailey, Bobby Amburn, Mike Pelfrey, Cindy Kovsky, Randy Loudermilk, Jan Beaube, Andrea Queen, and Rodney Arp.

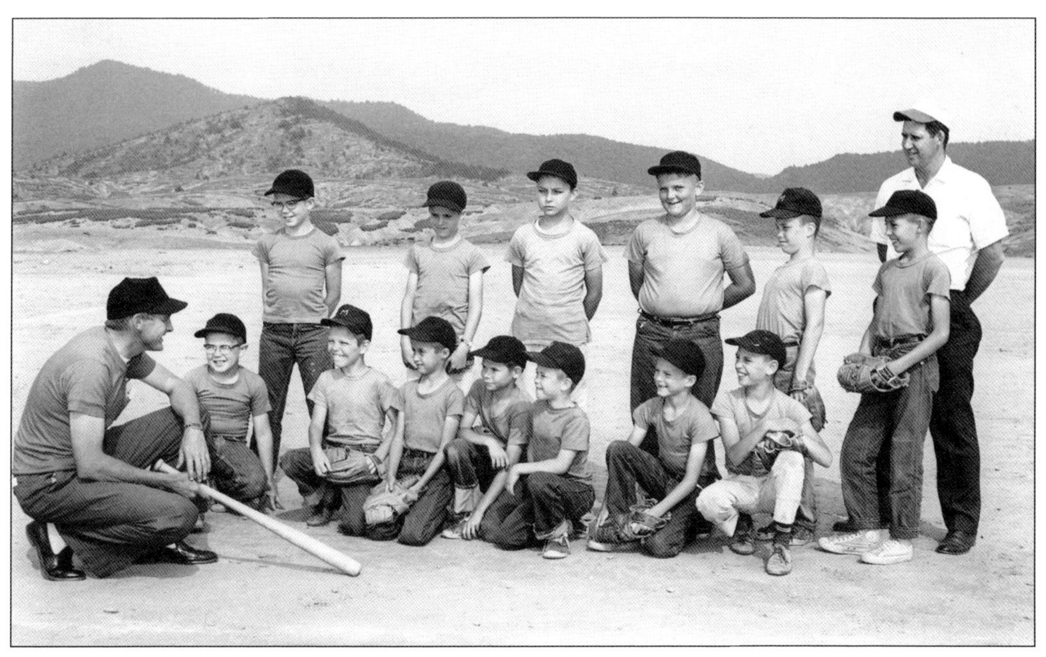

Dixie Youth Baseball, 1964. In summer 1964, ICWU Local 401 joined local civic clubs in sponsoring the Dixie Youth Baseball League, which consisted of ten teams of boys aged 8 to 12 and four "minor-league" teams of younger boys. Above, the Isabella Mets, led by manager Herbert Newman (left) and coach Jerry Sisson, get ready for a game. Below, on the old community baseball field near Burra Burra Mine, the Ducktown Cardinals listen to instructions from coach G.E. Fowler.

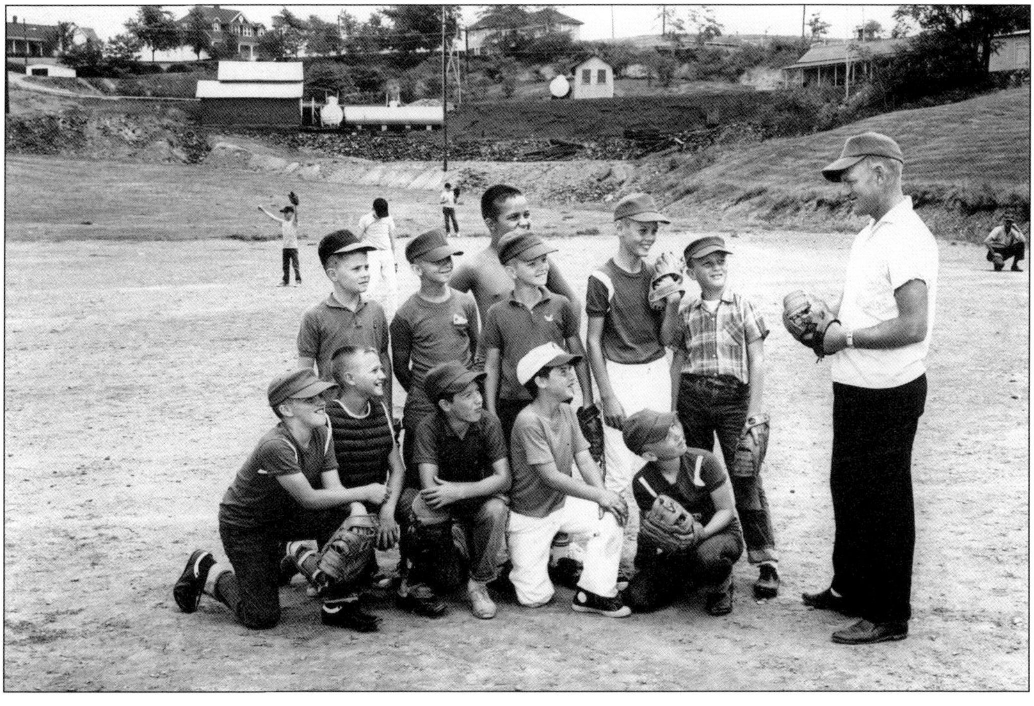

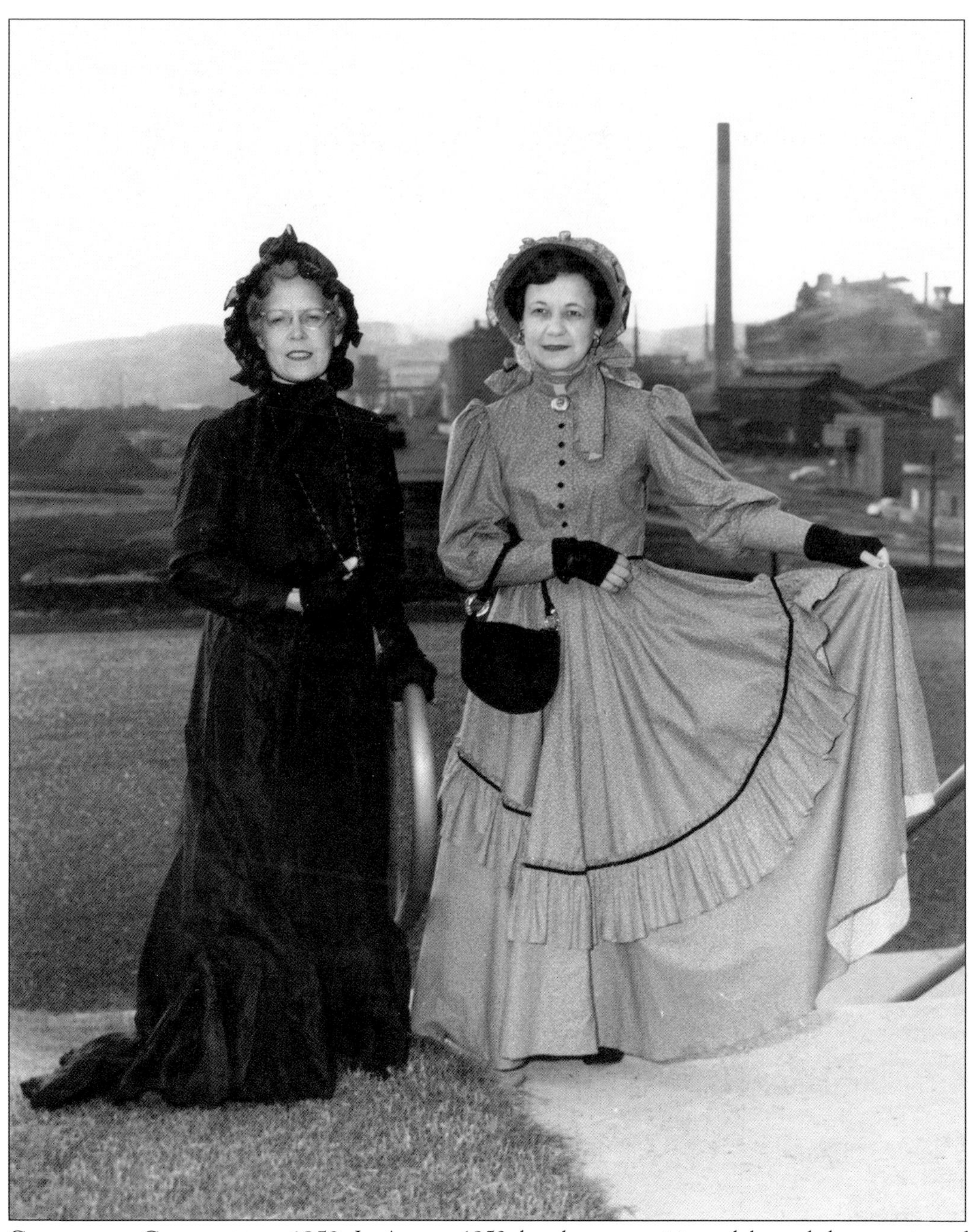

Centennial Celebration, 1953. In August 1953, local communities celebrated the opening of the first Ducktown district mines in a week-long Second Century celebration so popular it was repeated the following year. Many residents, including children, got into the spirit of the occasion by wearing period costumes, and men were required to grow beards or risk being captured and put into the "stockade" for noncompliance. In front of the TCC Copperhill plant, local schoolteachers Ida Belle Earnest (left), wife of supply department employee Zellie Earnest, and Marguerite German, wife of acid department employee Fred German, pose in their prize-winning costumes.

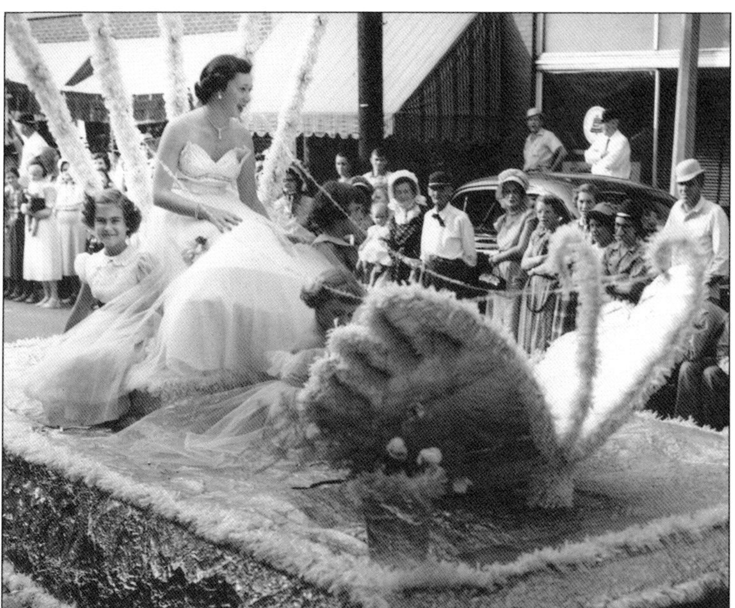

CENTENNIAL PARADE, 1953. On the Tennessee Copper Company float in downtown Copperhill, queen Mary Jo Norton is surrounded by attendants Judy Keffer (left), Betsy Hill (partly hidden by the butterfly), and Celie Burns. All four were the daughters of TCC employees.

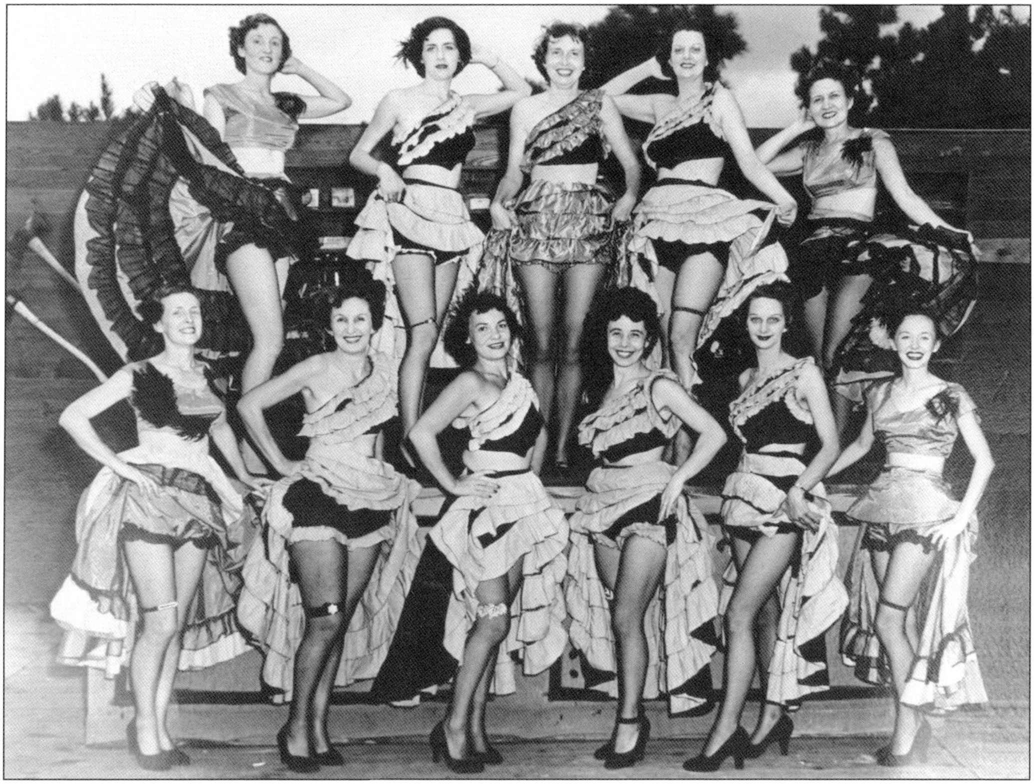

CAN-CAN GIRLS, 1954. *Copper Dust*, an original outdoor drama performed in a purpose-built amphitheater, was a highlight of the 1954 centennial celebration. At dress rehearsal for their dance routine are, from left to right, (first row) Louise Hawk, Ann Rankin, Shirley DeWitt, Titine Kopp, Peggy Brock, and JoLynn Simmons; (second row) Johnnie Weeks, Zane Adams, Sue Mitchell, Kathryn Gilliam, and Katherine "Cootie" Finch.

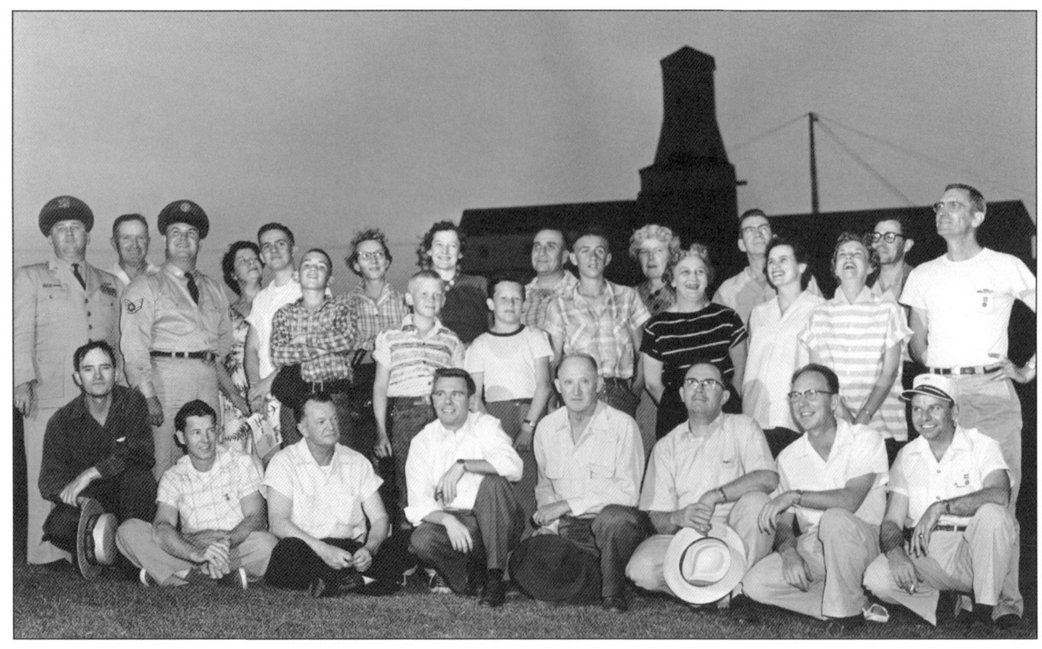

Civil Defense, Mid-1950s. During the Cold War era, the Copper Basin had an active Ground Observer Corps (GOC) as well as a large and well-supported Civil Air Patrol (CAP) unit with its own L-5 Stinson airplane and a Link aviation trainer. The photograph of GOC skywatchers above, with the Burra Burra Mine headframe in the background, was taken in May 1956 during the post's annual picnic and awards ceremony at the Ducktown ball field. Below, from left to right, CAP members C.L. Burnette, Robert Lohr, Gilbert Ratcliff, Haney Howell, Jon Earl Mason, Vic Hensley, and Bobby Dickey represent the patrol's mix of teenaged cadets and experienced pilots.

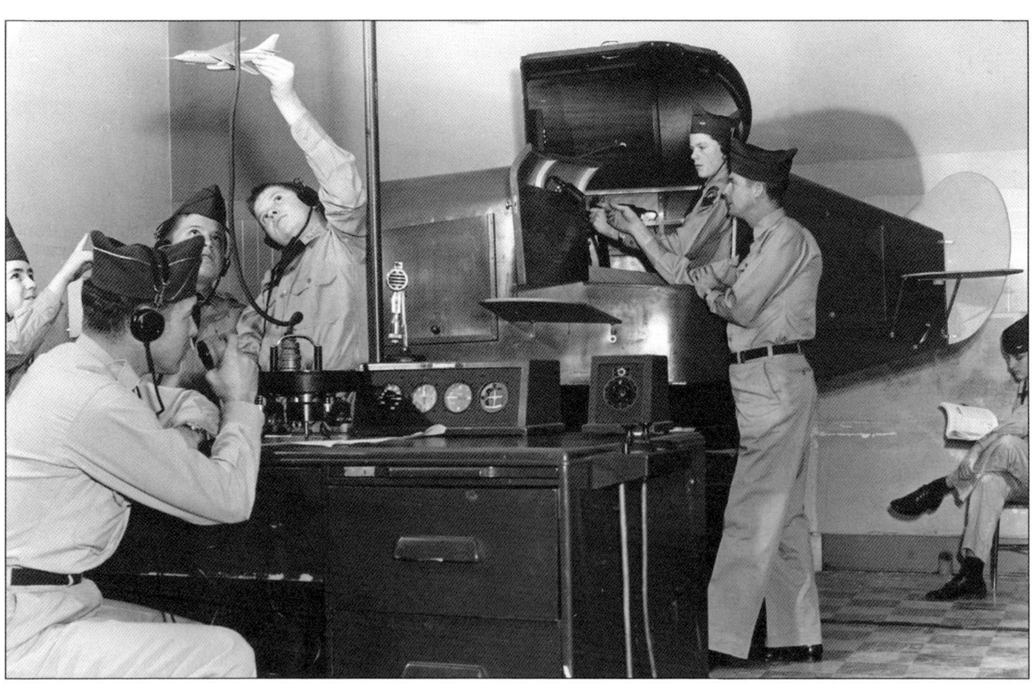

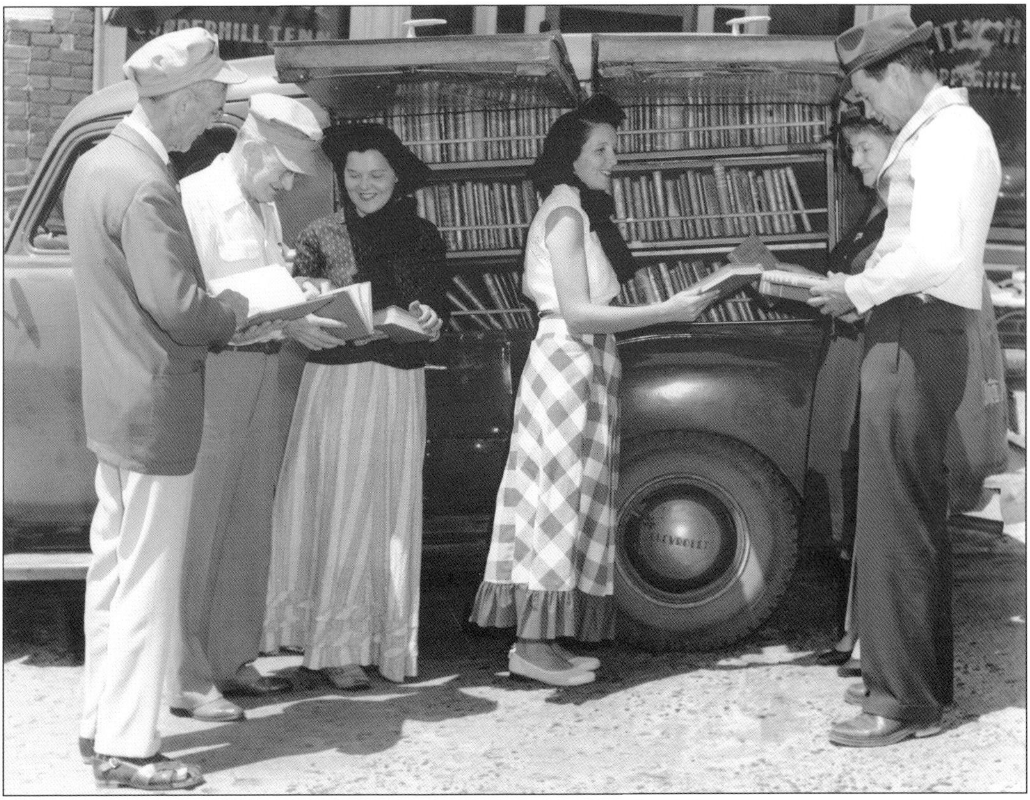

Bookmobile, 1953. In August 1953, during the centennial celebration, a new traveling library service was established at Copperhill City Hall. From left to right are McCaysville, Georgia, mayor Millard Cline, Copperhill mayor W.P. Terry, regional library employees Mary Melton and Evelyn Matthews, Fannin County, Georgia, library board chairman Mrs. J.C. Thomason, and Polk County library board chairman Floyd Kimsey.

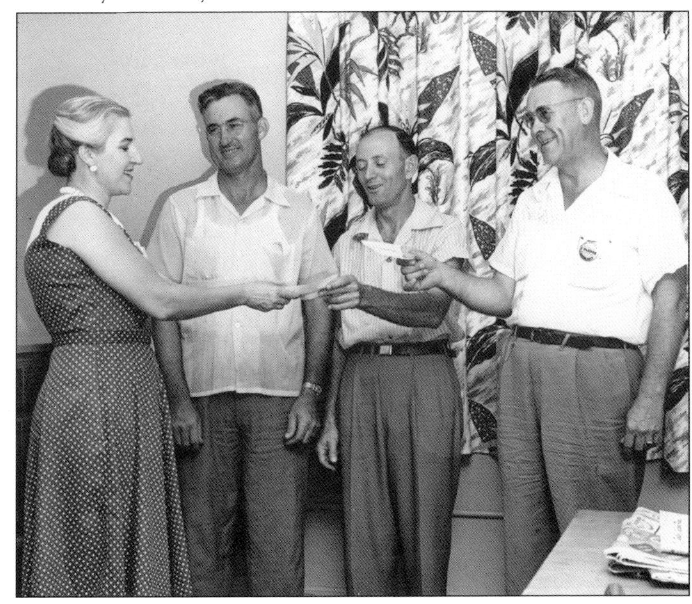

Library Opening, 1954. In summer 1954, a new library serving both Tennessee and Georgia residents was opened in the Copperhill YMCA building with contributions from civic clubs, TCC unions, and other groups. From left to right, Vera McConnell accepts donations from TCC union officials W.T. Phillips, R.H. Woodall, and Earl Hyde. The draperies were sewn by members of the Copperhill Iris Garden Club.

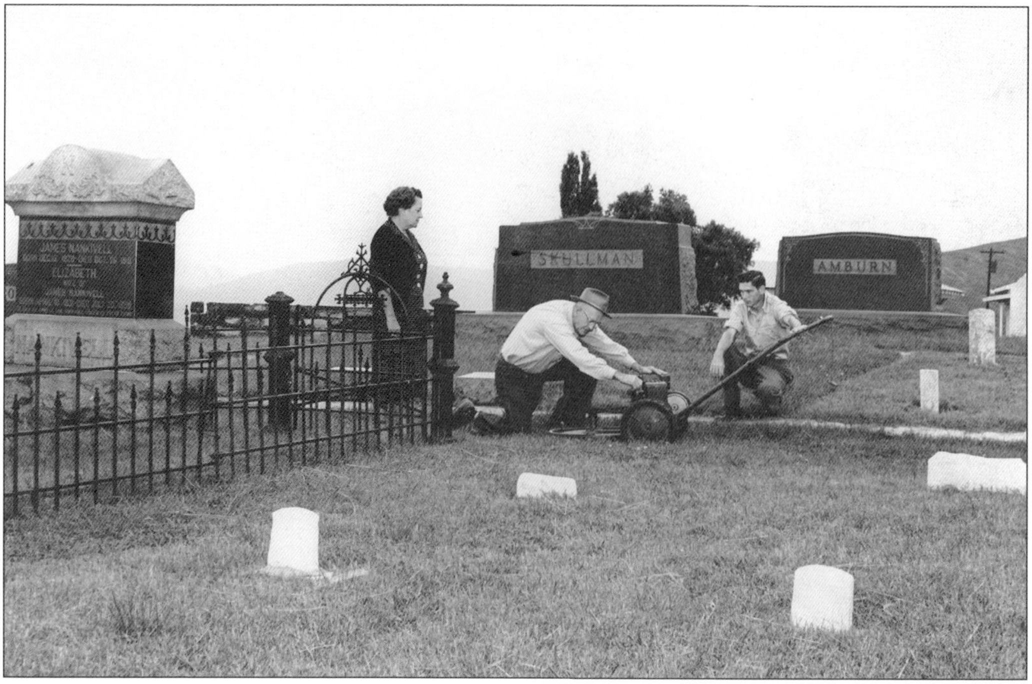

CEMETERY RESTORATION, 1960. In 1960, Vera Hyatt (left) spearheaded an effort to restore the century-old Ducktown Methodist Church cemetery, which had become overgrown with trees and vines. Hyatt was the wife of TCC medical director H.H. "Cooney" Hyatt and the daughter of early Burra Burra Mine captain C.V. "Cap" Skullman, whose family headstone is visible in the background. Assisting are project supervisor J.E. McMillon (center) and Wayne Ballew.

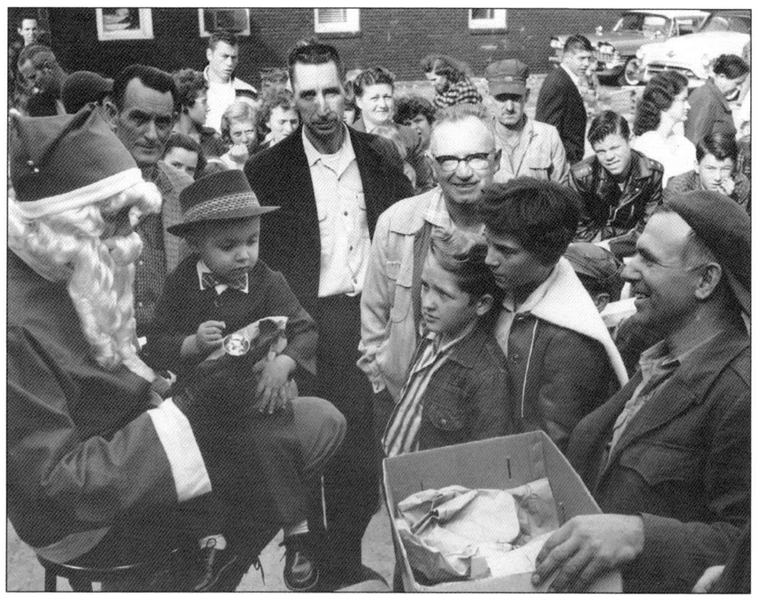

SANTA ARRIVES, 1960. On Christmas Eve 1960 in Copperhill, Santa Claus came to town in the person of TCC railroad worker Jack Housley, who gave out more than 1,000 bags of goodies to local children. On his knee is Jimmy Tallent. Gathered around, from left to right, are Young Pruitt, James Phillips, Norman Derreberry, and Roy Turner of ICWU Local 401, which co-sponsored the event.

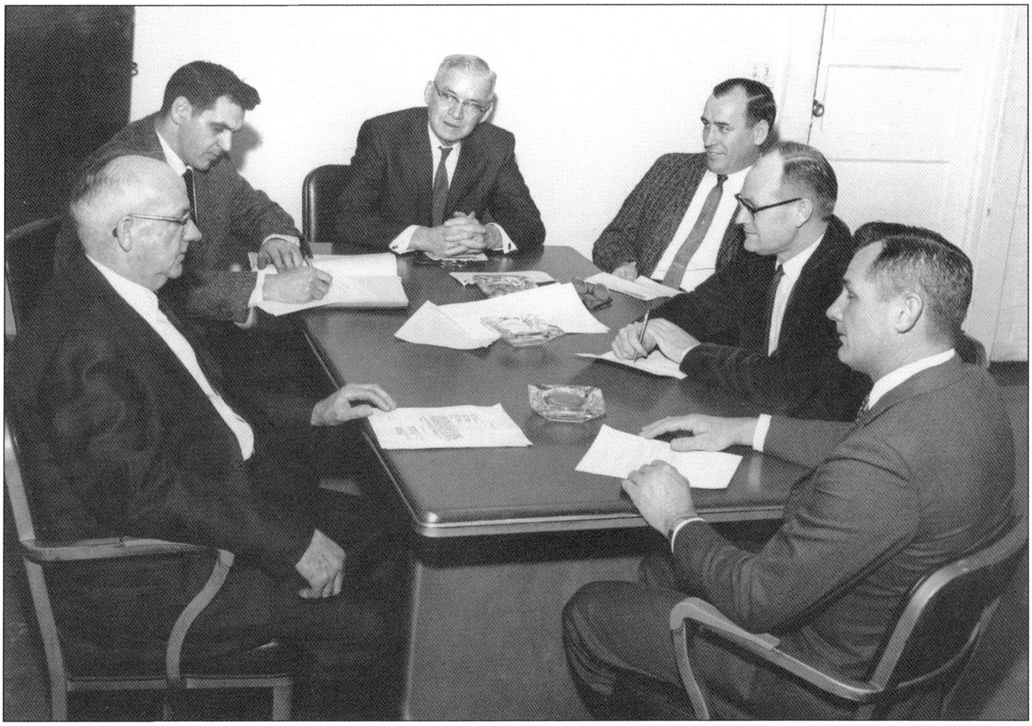

COPPERHILL CITY COUNCIL, 1965. Newly elected Copperhill mayor Frank G. Beaver (center), who had recently retired as TCC general carpenter foreman, chairs a meeting of the Copperhill City Council. From left to right are councilmen Carl Arp and Bill Tallent, Beaver, and councilmen Bob Kilpatrick, E.V. Brackett, and Edward Swanson. Arp was the only councilman who was not a TCC employee.

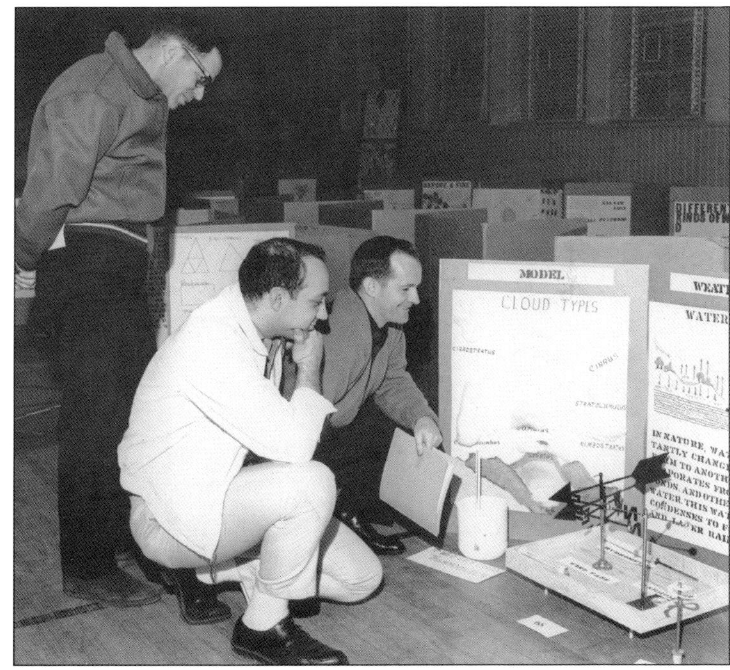

SCIENCE FAIR, 1964. Beginning in the late 1950s, TCC engineers were leaders in helping to stage annual science fairs at local elementary and high schools. Other TCC employees also took active parts in the program. In this 1964 photograph, from left to right, George DeWitt, Copper Basin High School teacher Ferris Maloof, and Robert Cash judge a student display at Turtletown Elementary School.

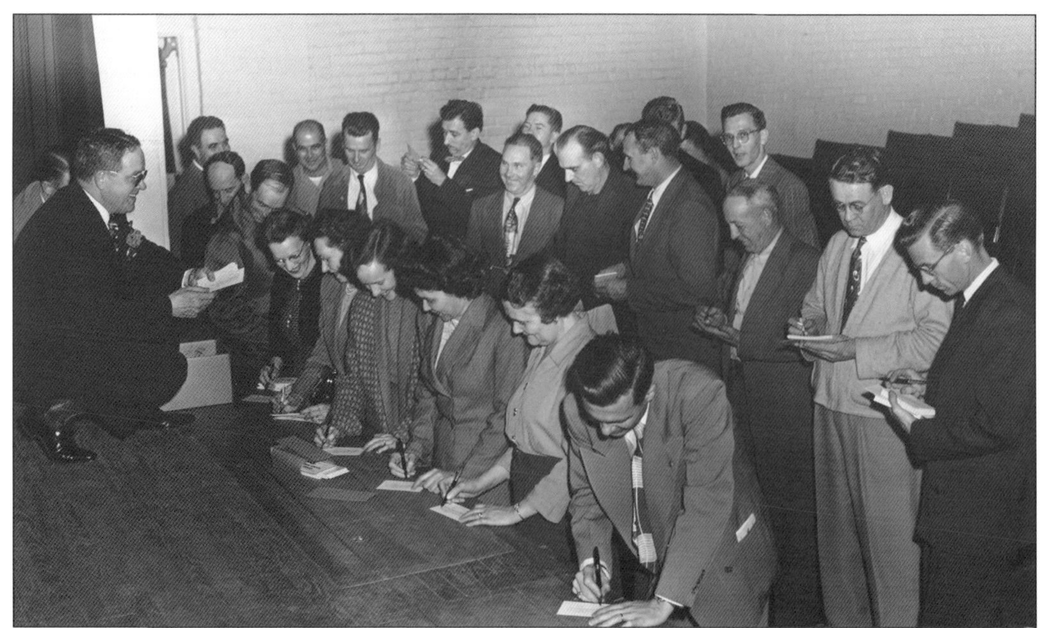

New Community Hospital, 1950s. On February 2, 1953 (above), TCC employees were among the group of local residents and business owners who gathered for the kickoff of a fund drive to build a modern community hospital. Federal and state funds covered two-thirds of the costs, but the remaining $245,000 had to be raised locally. By mid-March, as their part of the community-wide fundraising effort, nearly 1,500 TCC employees had pledged a total of $77,000. TCC contributed an additional $50,000 and provided land, equipment, and operators for the preparation of the site. In August 1955, the Copper Basin General Hospital opened its doors with TCC medical director H.H. "Cooney" Hyatt (below, with an unidentified patient) as chief of staff.

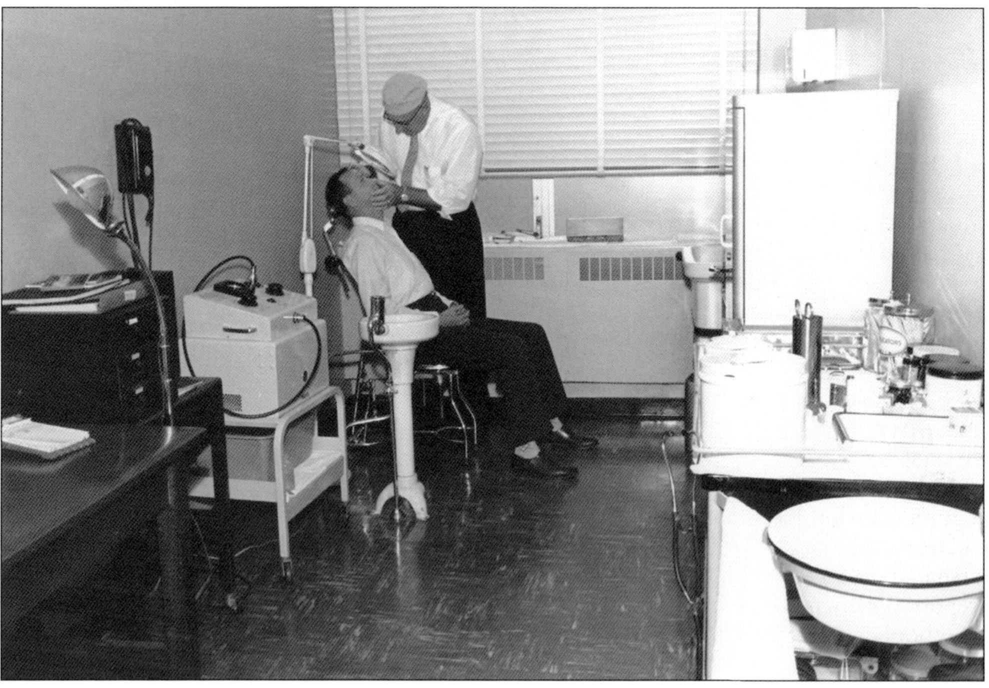

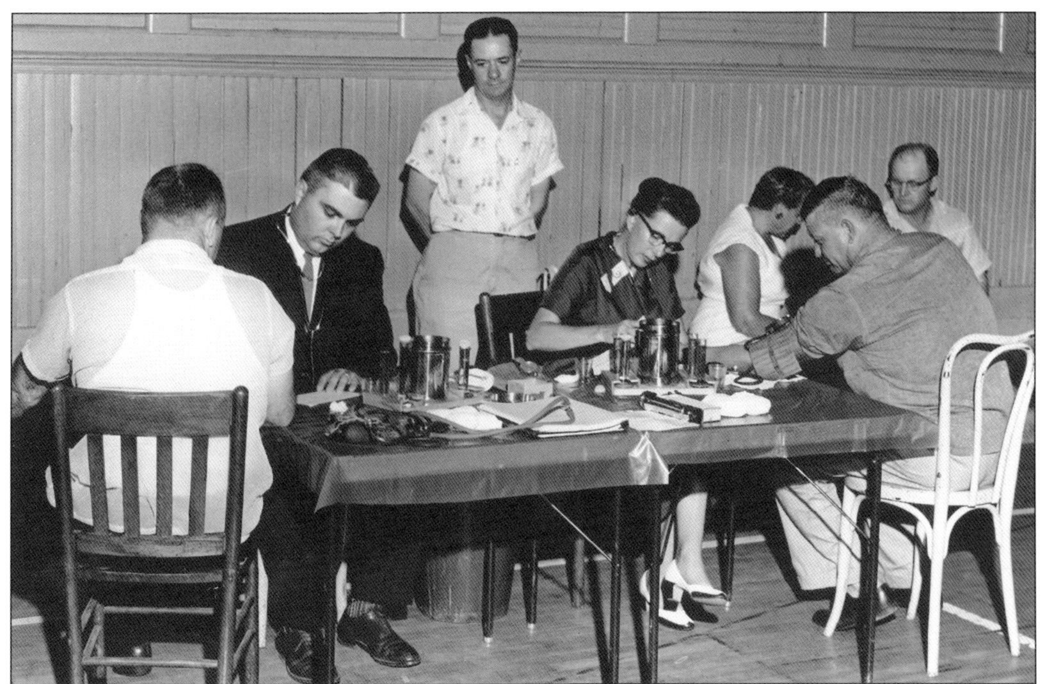

Bloodmobile Visit, 1957. On the Red Cross bloodmobile's quarterly visit in June 1957, TCC employees and other area residents donated 95 pints of blood. At the doctor's table, Dr. J.T. Layne (second from left) is assisted by registered nurses (and employee wives) Madelyn Kendall (fourth from left) and Genevieve Denman. Standing is blood bank committee chairman Glenn Thurman. Being checked before donation are Eckord Foster (far left), Homer Ricord (second from right), and L.J. Wise.

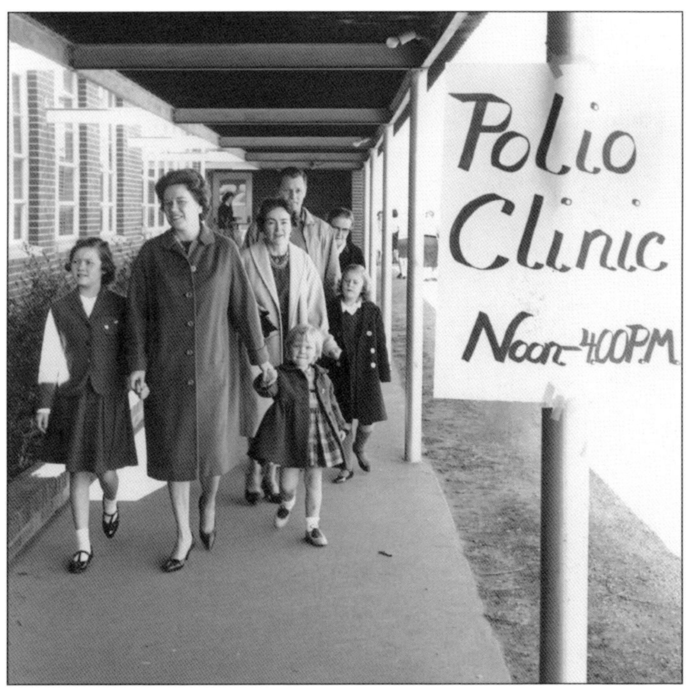

Sabin Oral Sunday, 1964. On February 2, 1964, more than 10,000 local Tennessee and Georgia residents participated in a polio vaccination clinic sponsored by Copper Basin General Hospital. In front, at the Copper Basin High School vaccination site, are Cathryn McCarter with daughters Joetta (left) and Carol; behind them are Eunice Taylor, Joe McCarter with daughter Marsha, and Lula Quinn. A reported 91 percent of area school children were vaccinated that day.

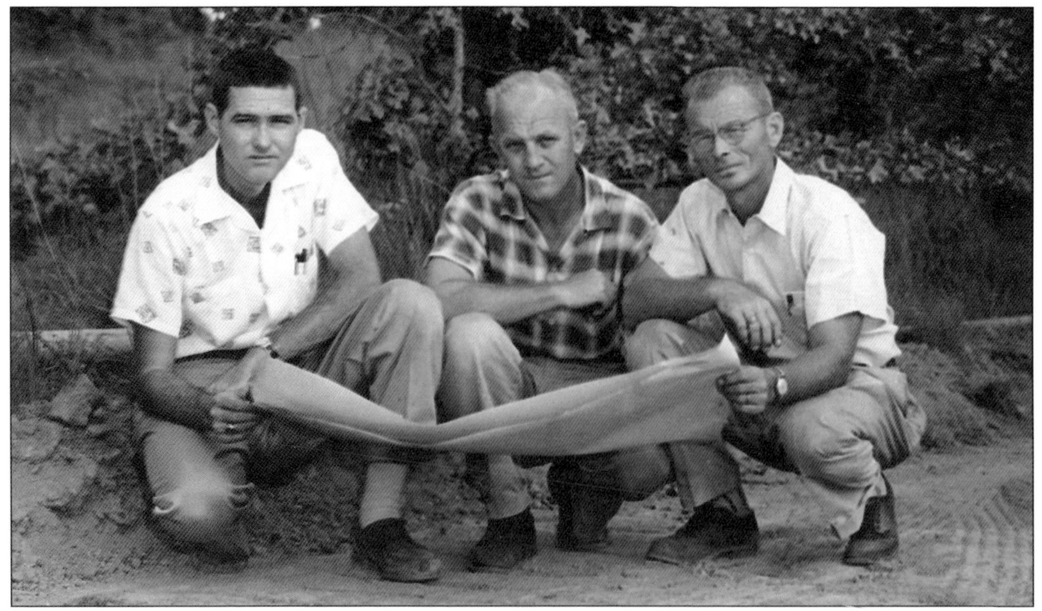

Cherokee Hills, 1957. In the mid-1950s, TCC offered 34 lots in the newly surveyed Cherokee Hills neighborhood, just outside Ducktown, for sale to employees at nominal prices. Construction and infrastructure were the responsibility of the homeowners. Above, from left to right, Cherokee Hills Utility District commissioners Bob Lynch, John Rankin, and Fred Panter pose with plans for extension of the water system. The new Panter home, below, was the second one occupied. Helping decorate their new living room, from left to right, are Glenda Sue, Fred, Fredda, and Gladys Panter.

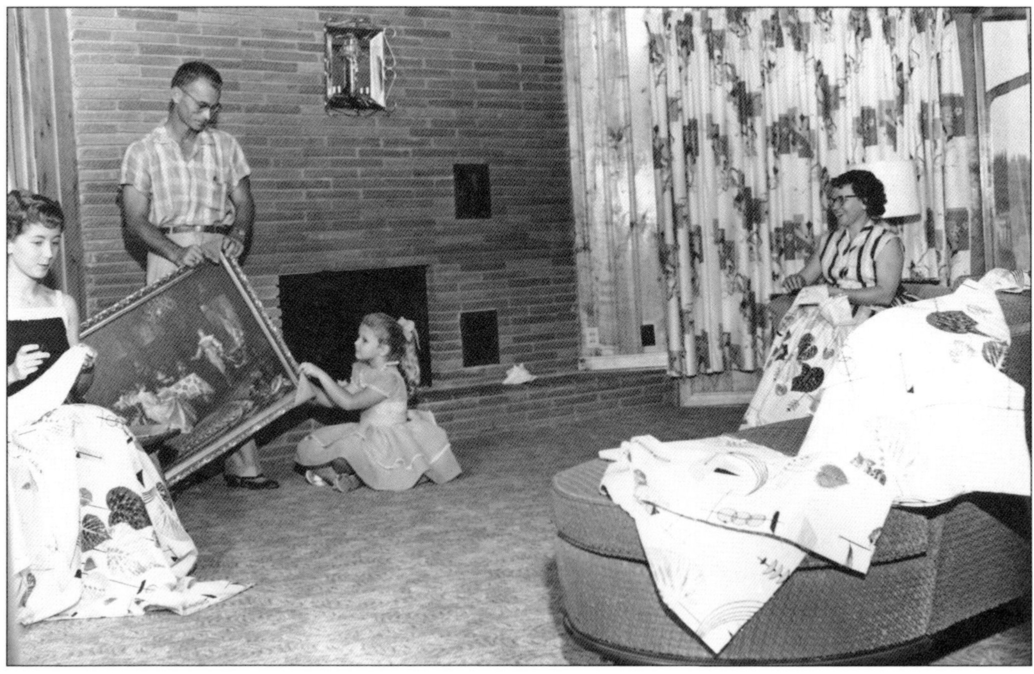

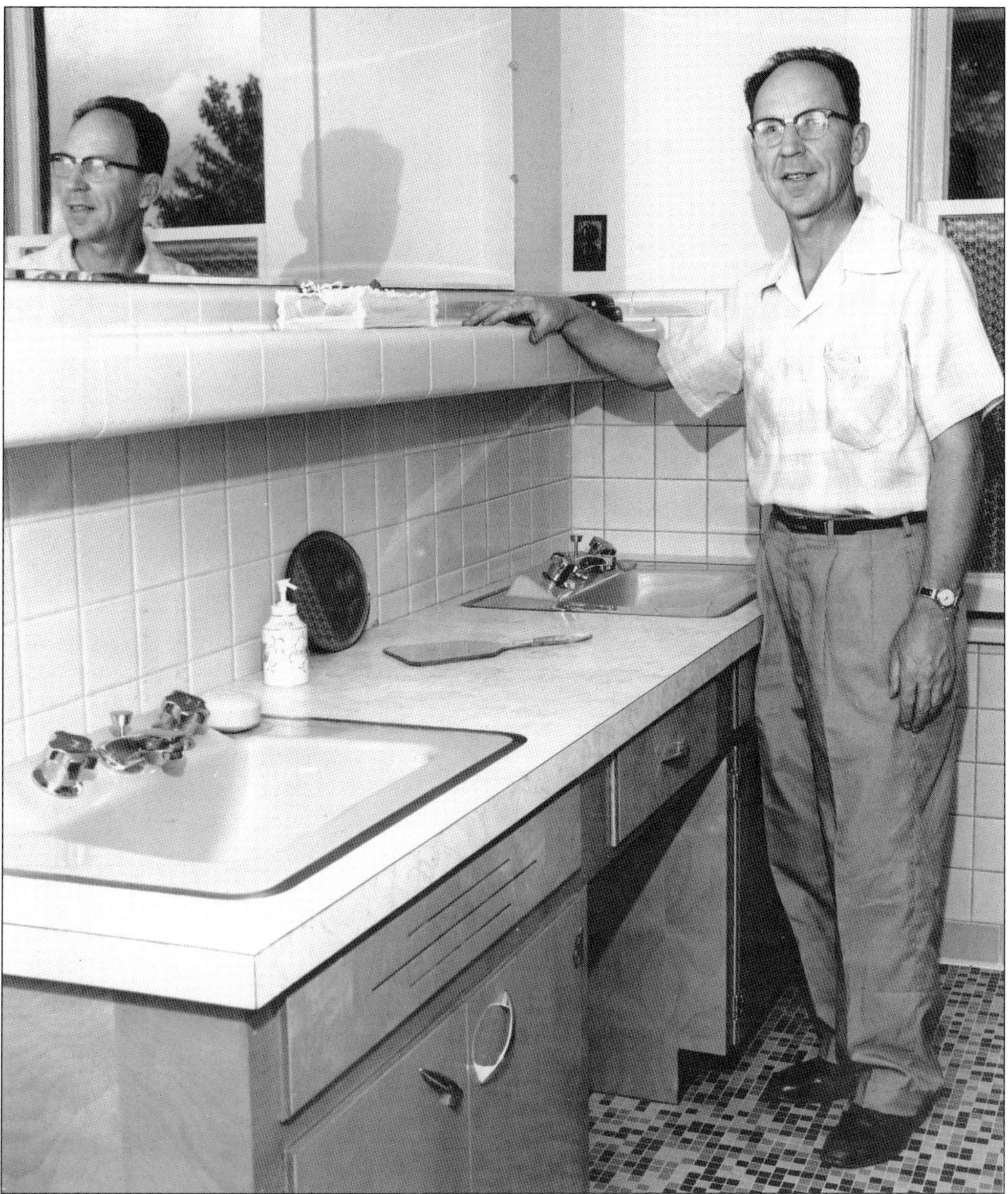

From Hospital to Home, 1962. With the opening of Copper Basin General Hospital in 1955, both the TCC company hospital in Copperhill and the Kimsey-Guinn hospital in Ducktown were permanently closed. Seven years later, after having lived in company rental housing for 15 years, mining engineer Guerdon Kopp and his wife, Titine, bought the now-empty 1928 Kimsey-Guinn building in Ducktown and set about turning it into a residence. The former operating room became a family room, the nursery became the dining area, and the emergency and delivery room was converted into the large bathroom in this photograph, with its original ceramic tile preserved. Kopp's right hand rests on a shelf that once held small medical instruments.

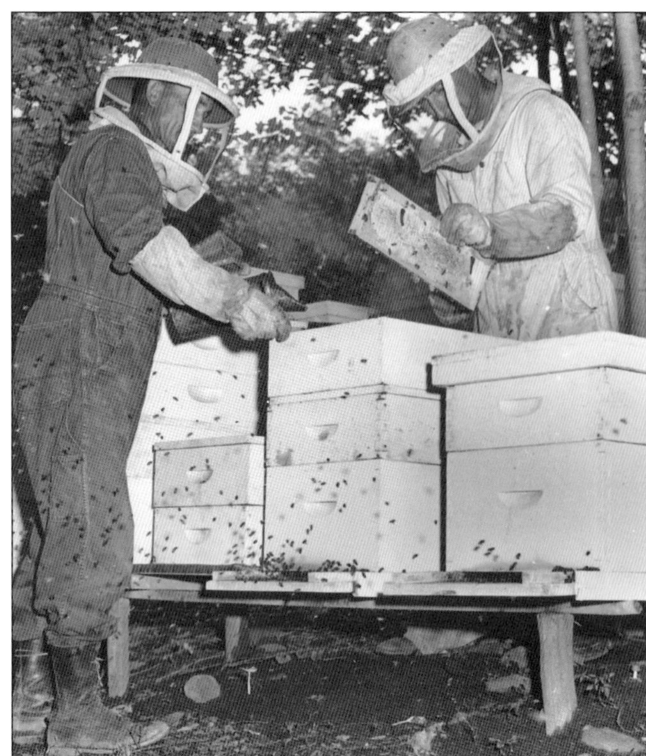

Beekeeping, 1957. According to accounting department employee Keith Postelle (right), beekeeping was a favorite hobby of many of his TCC coworkers. Postelle and partner Henry Smith developed their hobby into a small business. They maintained about 90 hives, extracting and packaging the honey for local sale. Every two years, they introduced new queen bees in order to "re-queen" the colonies.

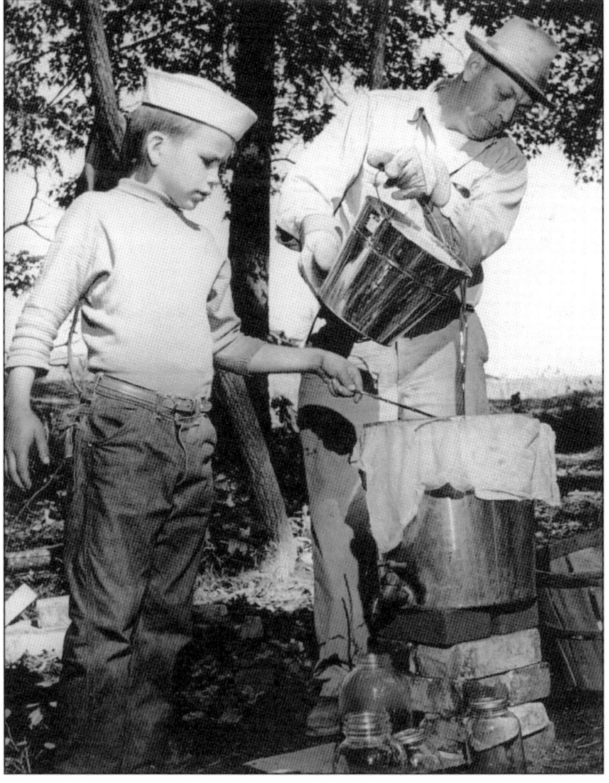

Sorghum Making, 1961. In 1961, mine repairman C.H. Gibbs decided he wanted to make sorghum the way his father used to make it. The project involved the summer planting of cane, the building of a mill to extract the juice, and the construction of a furnace and evaporator for cooking down the liquid. In the next to last step, aided by his son Dewey, Gibbs pours the sorghum into containers for cooling.

Apple Picking, 1963. In the Ducktown yard of Isabella acid department employee Ervin King, overlooked by the Burra Burra headframe, was a Red Delicious apple tree that turned 25 years old in 1963. It also had a particularly good season that year—so good that its lower branches required propping up. King had owned the house for two years and had spent considerable time improving the yard with rich mountain soil.

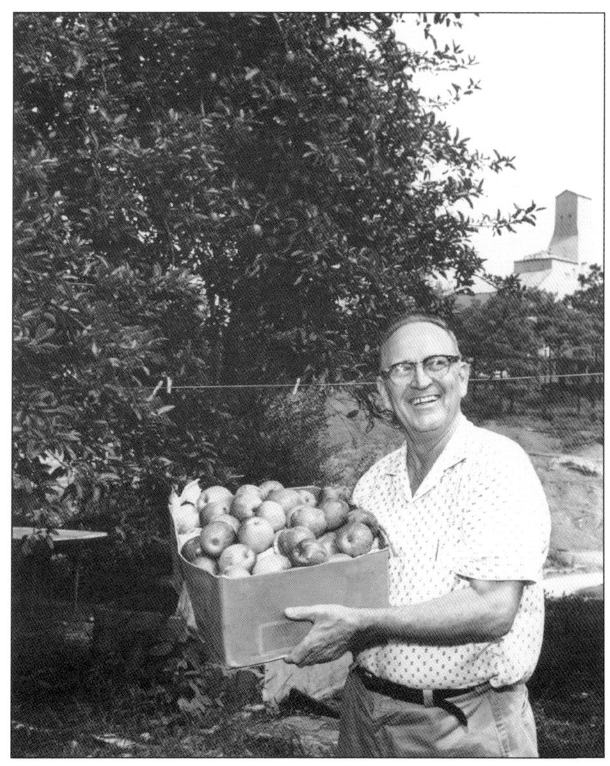

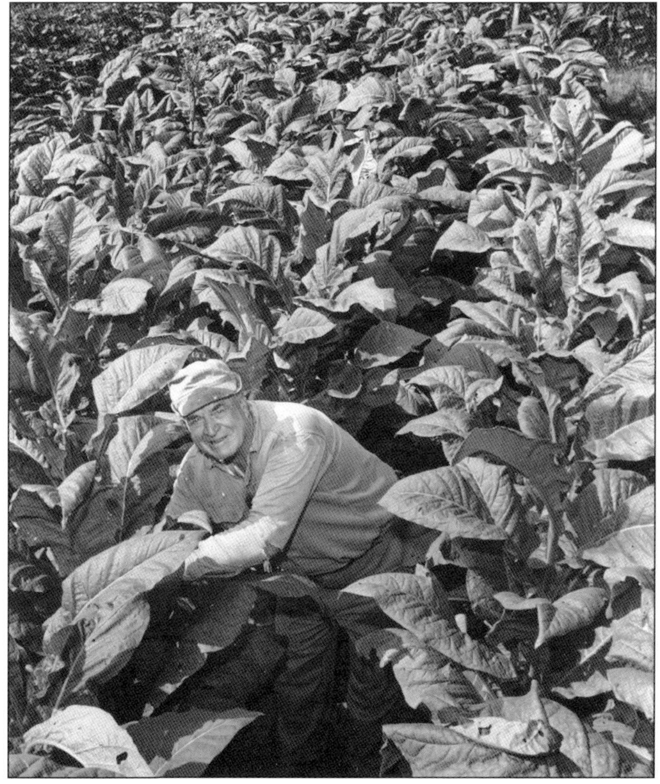

Home-Grown Tobacco, 1966. In retirement, Eureka miner J.H. Keener found more time to enjoy his latest aboveground occupation: growing his own tobacco. It was never meant to be a commercial enterprise; he planted just a few rows every summer, hung his harvest to dry, saved seed for the next planting, and crushed a handful of leaves when his pipe needed filling. The result, he said, was "a real mellow smoke."

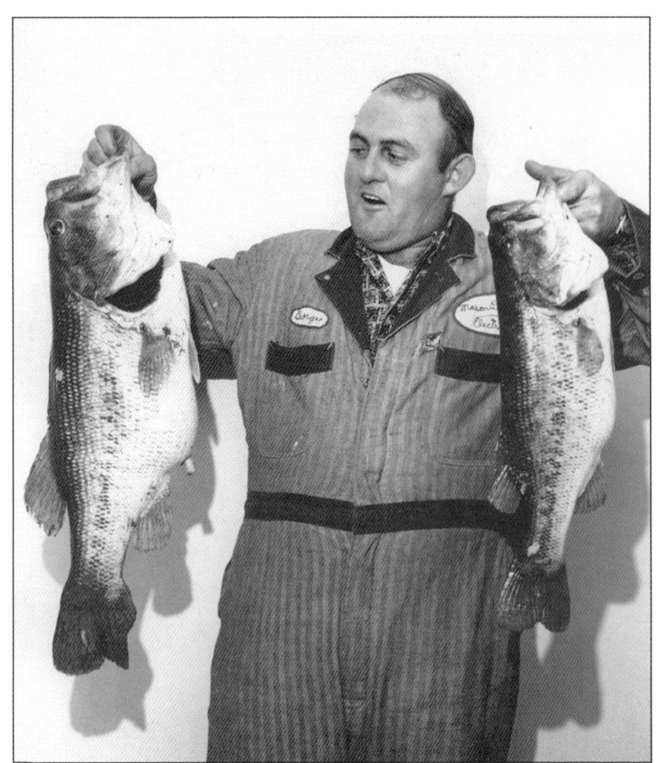

FISH STORY, 1962. In December 1962, acid plant employees Gene Wade Whittenbarger and Clint Carter came home from a fishing trip with 18 pikes and bass, including these two trophy-sized largemouth bass landed by Whittenbarger. The biggest bass weighed 10 pounds, 9 ounces, and was 27.5 inches long. Whittenbarger reportedly went fishing again the next week and came home with a catch of 24 more.

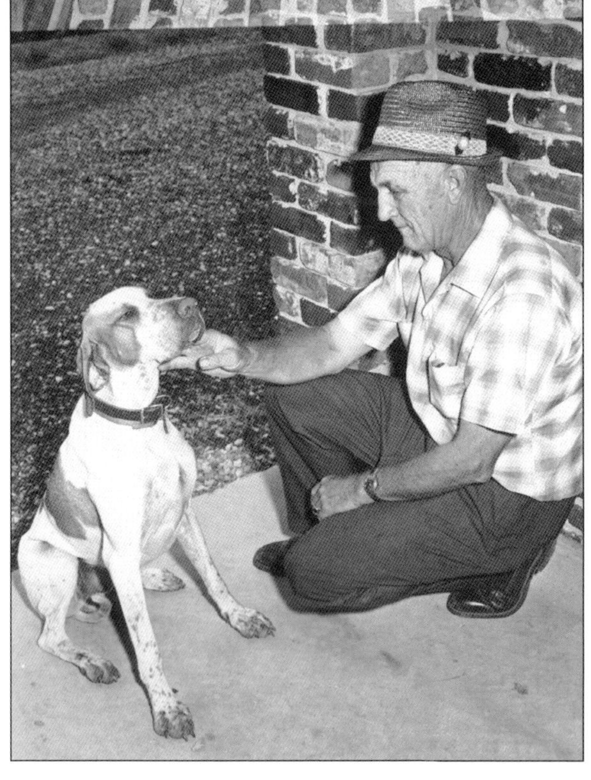

PRIZE PUP, 1971. John Rankin of the TCC engineering department never intended Sergeant Sam to be anything but a family pet. Then his daughter Libby Barclay talked him into entering the nine-month-old pedigreed English pointer in the Atlanta dog show. Sam swept the puppy classification, was named best of show, and went on to win ribbons throughout the Southeast. Sam, according to Rankin, remained unimpressed.

Local Arrowheads, 1965. Retired power and electric foreman E.E. Hellerstedt displays the collection of arrowheads gathered by his family over the years around Ducktown and Isabella. Most of the arrowheads were thought to have been crafted by the Cherokees, and some had been donated to the Smithsonian Institute. Hellerstedt holds his prized collection of stamps, which dated to his boyhood.

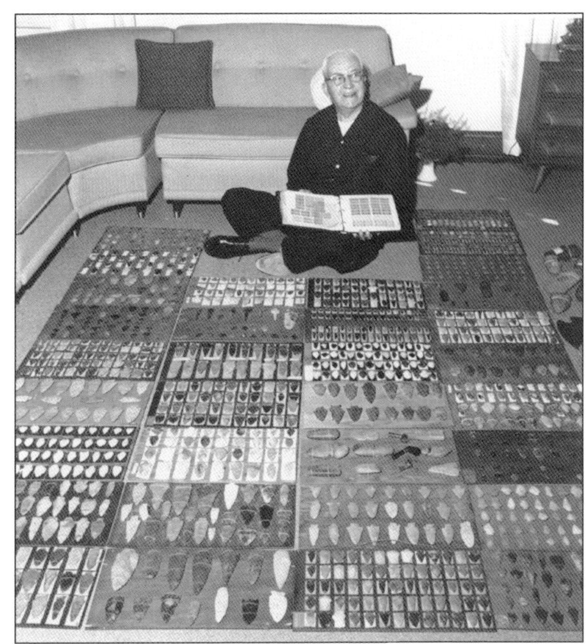

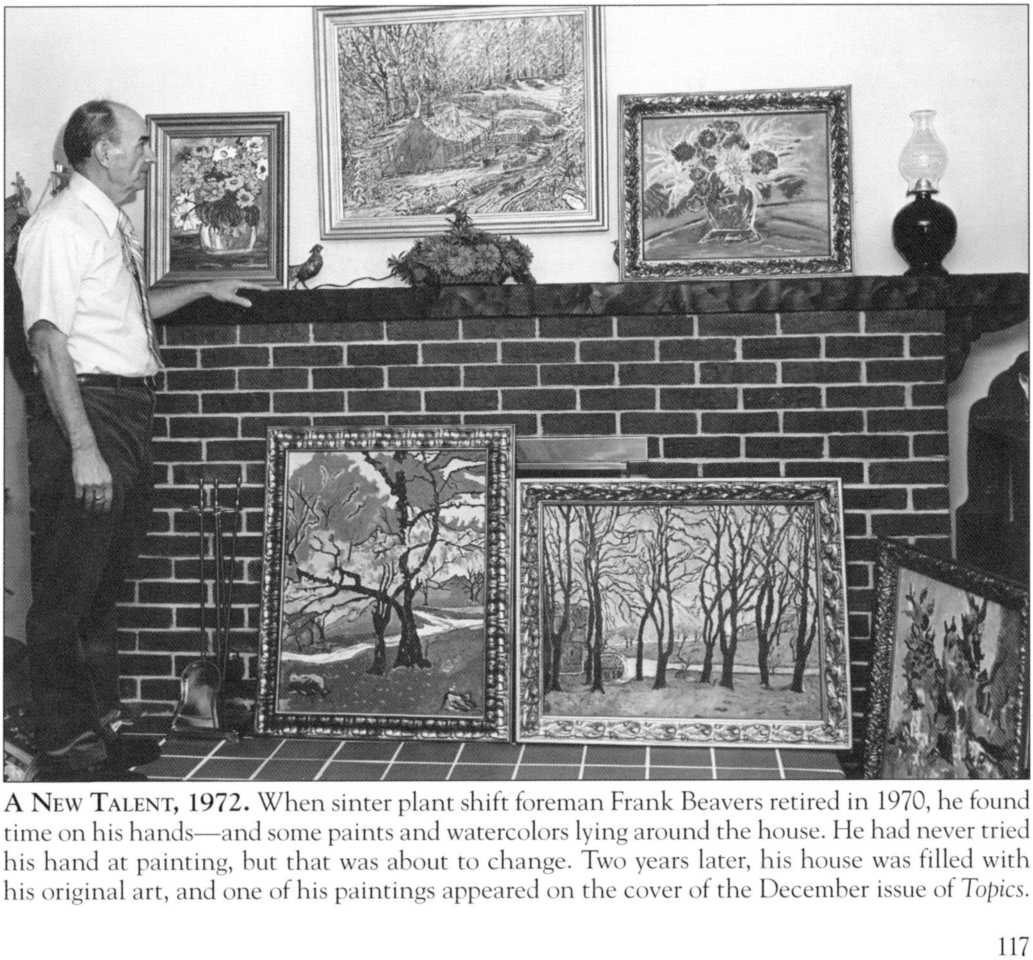

A New Talent, 1972. When sinter plant shift foreman Frank Beavers retired in 1970, he found time on his hands—and some paints and watercolors lying around the house. He had never tried his hand at painting, but that was about to change. Two years later, his house was filled with his original art, and one of his paintings appeared on the cover of the December issue of *Topics*.

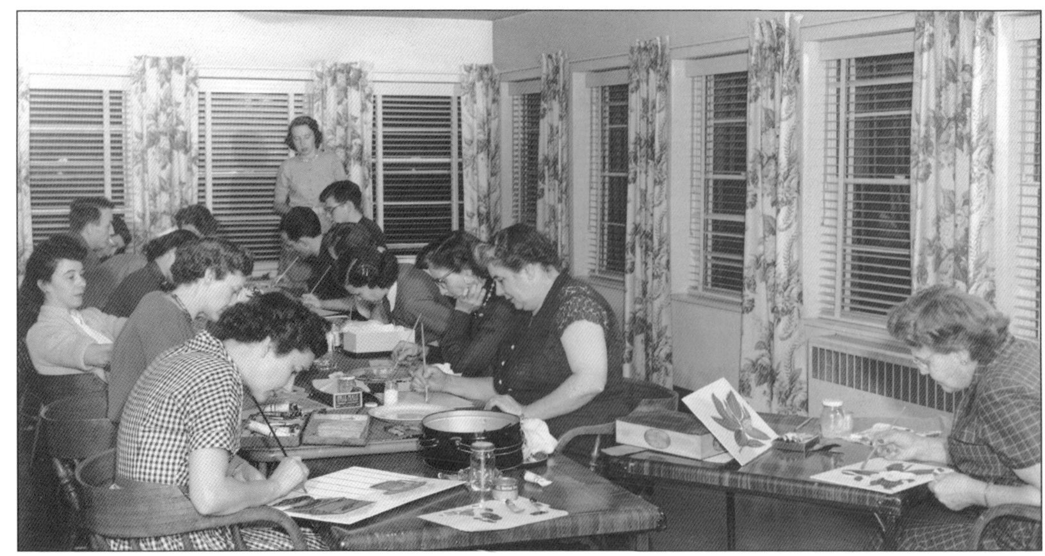

Art Classes, 1953. In November 1953, by popular demand, nationally recognized artist and Copperhill native Sue Mitchell agreed to teach a series of local art classes for adults. The classes attracted about 25 participants, most of them TCC employees or family members. Mitchell was the daughter of TCC general manager T.A. Mitchell.

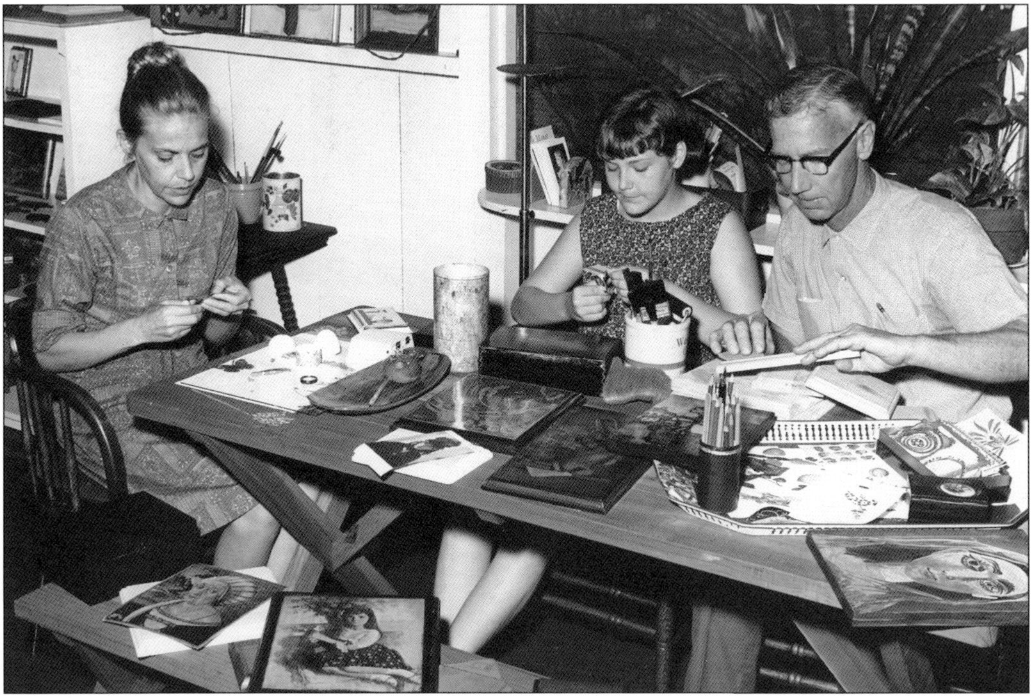

Art in the Family, 1966. In the home of supply department employee Oren Postelle and his wife, Grace, art was everywhere. Grace delighted in exploring new art forms and frequently got the whole family involved. In this August 1966 photograph, the Postelles and their daughter Becky work on a decoupage project, with Becky cutting out paper shapes and Oren responsible for the hand sanding of boards.

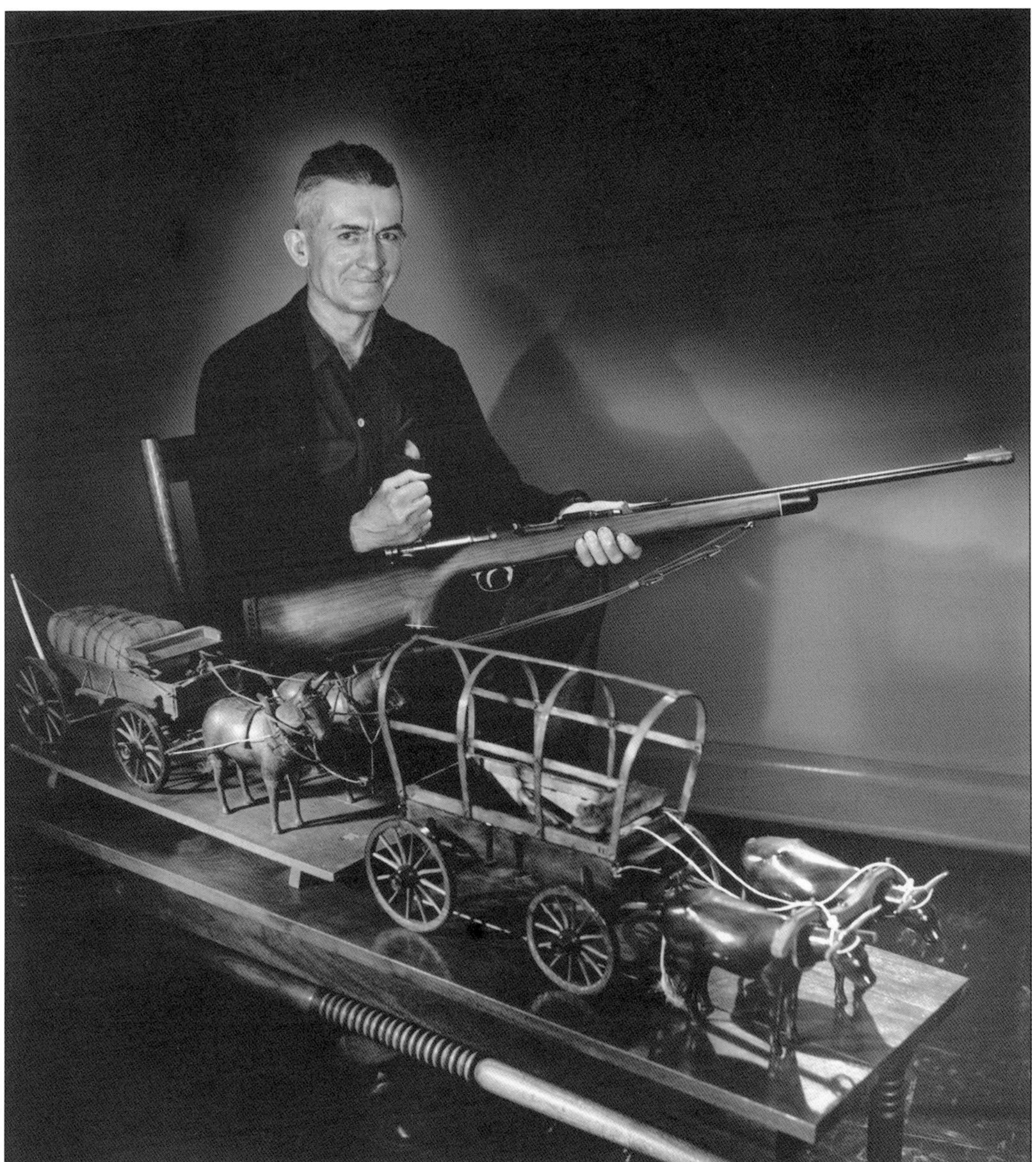

THE ART OF THE CARVING KNIFE, 1952. On a winter day in 1928, sinter plant repairman Carl Stepp sat down behind the stove in his house to get warm. "For lack of anything in particular to do," he later recalled, "I took out my pocket knife and started to whittle." It was the inauspicious beginning of a hobby that would lead to his recognition as an accomplished folk artist whose intricate carvings included objects such as the horses, oxen, and wagons pictured here. Stepp also made the rifle he is holding. Some company executives received ashtrays designed with a small, meticulously carved acid tank car on its own railroad track, and Stepp once reportedly carved an ashtray with a donkey and sent it as a gift to Franklin Delano Roosevelt. He branched out into making furniture, a talent he put to use as a Seabee in World War II. Several examples of his work are in the permanent folklife collection at the Atlanta History Center.

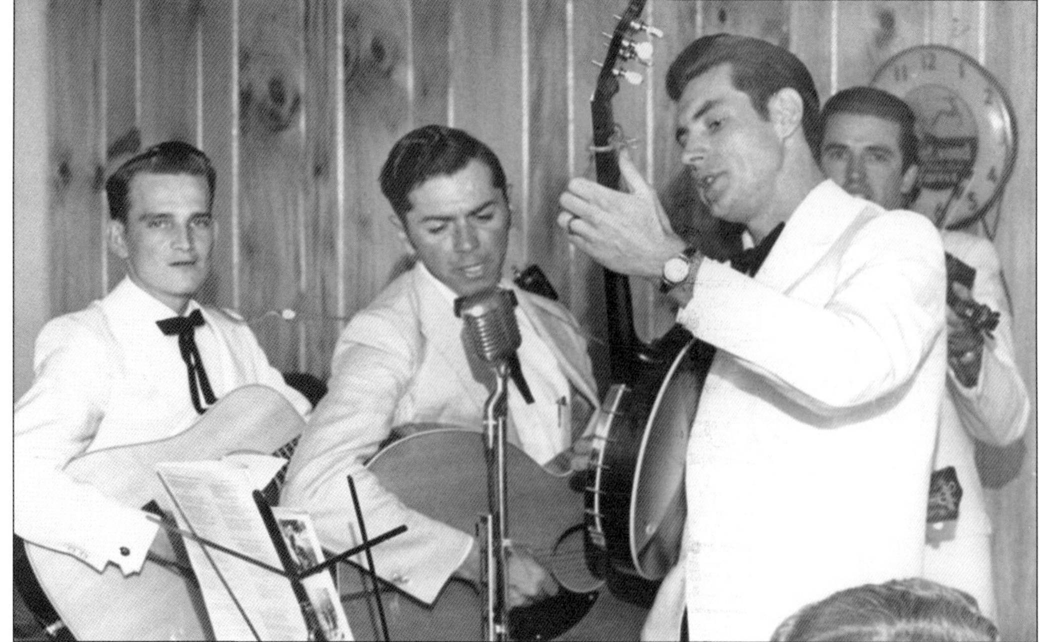

The Georgia Mountain Boys. In 1958, when the Georgia Mountain Boys were making regular radio appearances on Saturdays, three of the four band members pictured above were spending their workweeks at TCC: Ulas Thomas (far left) in the plate shop, Gerald Heaton (on banjo) at the Copperhill smelter, and Johnny Jones (rear) in the acid department. (At the microphone is Herbert Legg.) By January 1965, when the band released its first self-titled LP, its members (seen below from left to right) consisted of supply department employee Earl Dehart, Jones, Benny Collins (the only non-TCC employee), acid department employee Cleve Oliver, Heaton, and copper sulfate plant employee Wayne Walker.

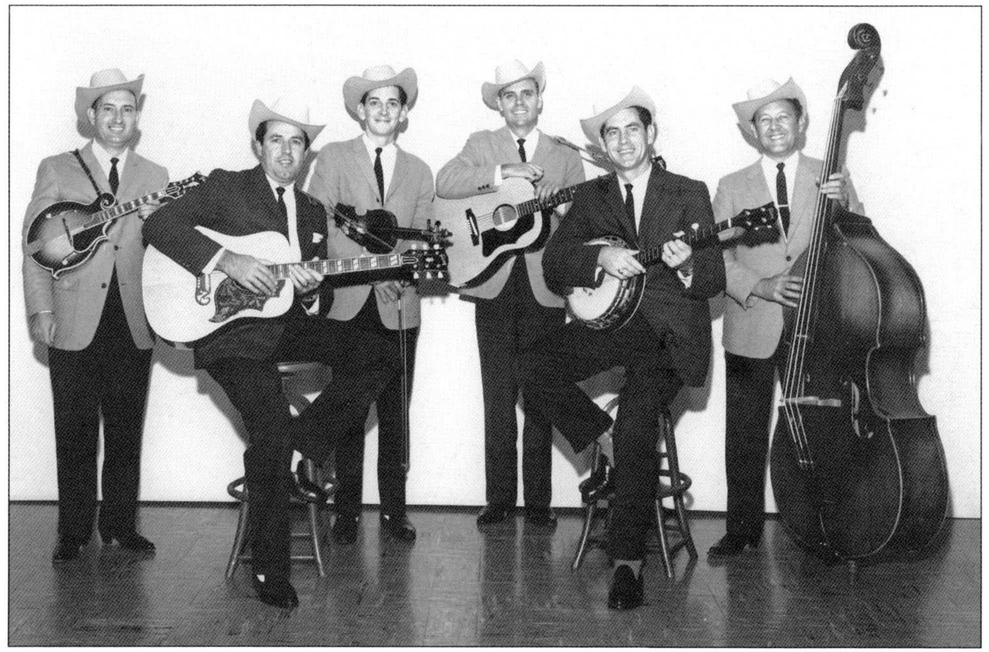

Mining Memories, 1953. Among the prized possessions of retired Burra Burra carpenter shop employee W.A. "Lon" Carver was this 19th-century iron candlestick he found years earlier in an abandoned drift at the mine The old candlesticks had sharp spiked points that could be jammed into mine timbers or cracks in the wall; the hook and loop handle made them easier to carry and fasten.

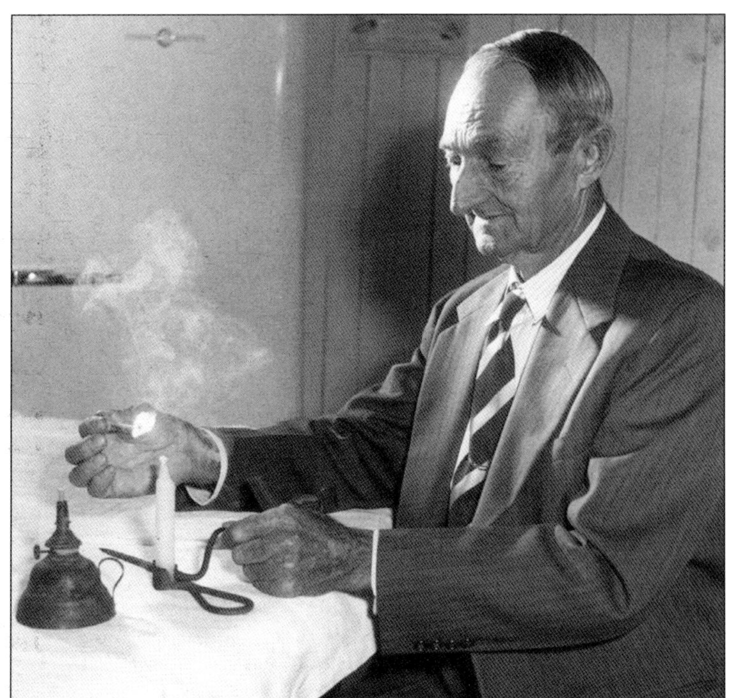

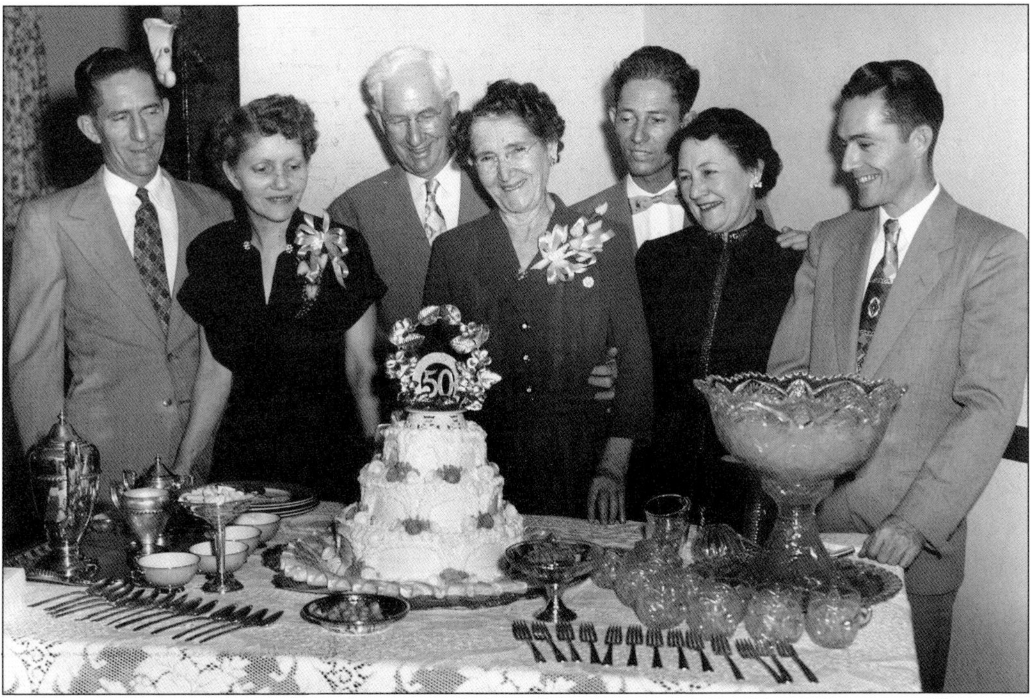

A TCC Family, 1957. Surrounding retired plate shop foreman C.P. "Ches" Queen and his wife, Nora, at their golden wedding anniversary celebration are their children, from left to right, Cyril, Lois Jones, Rupert, Ruby Amburn, and Charles. Queen had worked at TCC for 47 years when he retired in 1952. Both Cyril and Rupert set long service records of their own, and both Lois and Ruby were married to long-serving TCC employees.

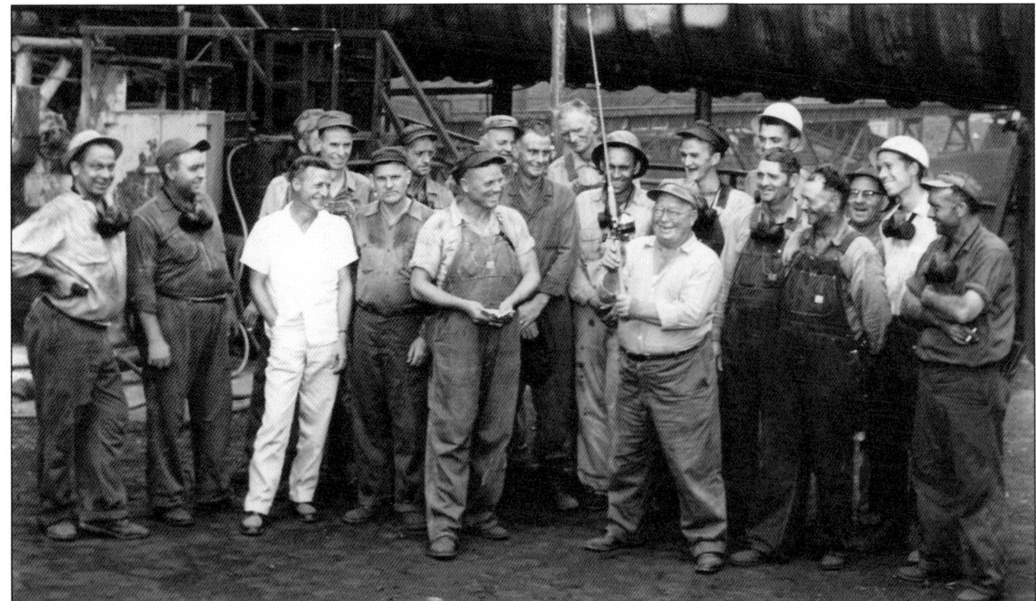

Retirement at the Roasters, 1961. Retirements at TCC were almost always an occasion for a celebration, and usually for a photograph that would run in the next edition of *TC Topics*. This celebration for roaster crewman Otto L. Ross, retiring to his farm after nearly 35 years on the job, was typical of the gatherings, which included retirement gifts such as the new rod and reel in Ross's hands.

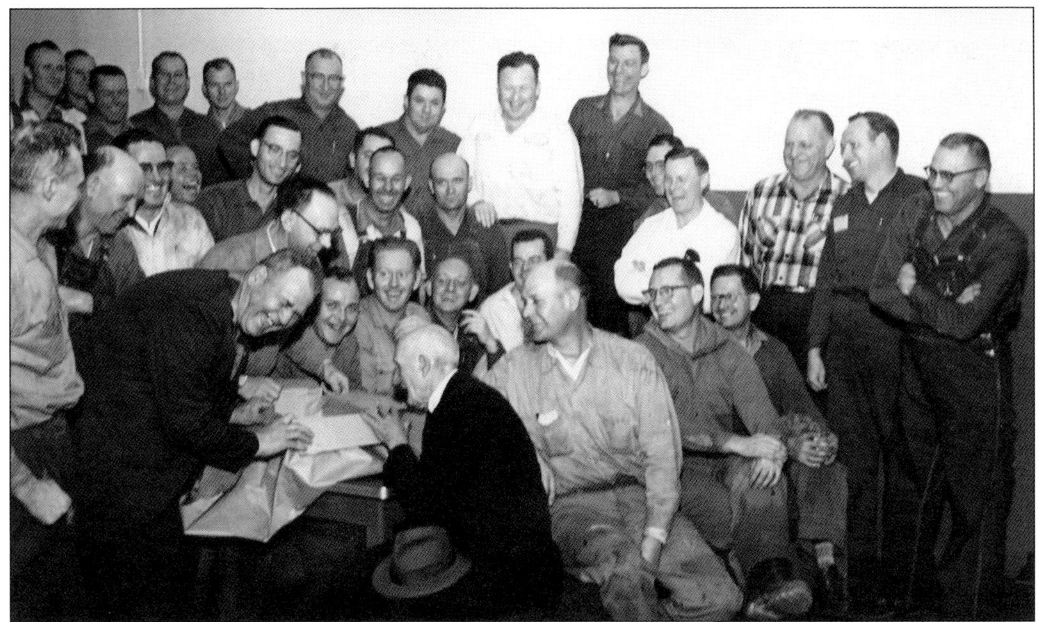

Retirement at the Acid Plant, 1963. In 1926, C.T "Tiny" Swafford was actively recruited by TCC—for his skill with a baseball bat. He joined the company-sponsored semiprofessional team that year and promptly became its star hitter. When he retired 37 years later from his official job as an acid repair foreman, he was apparently unprepared for his retirement gift: a neatly wrapped miniature still.

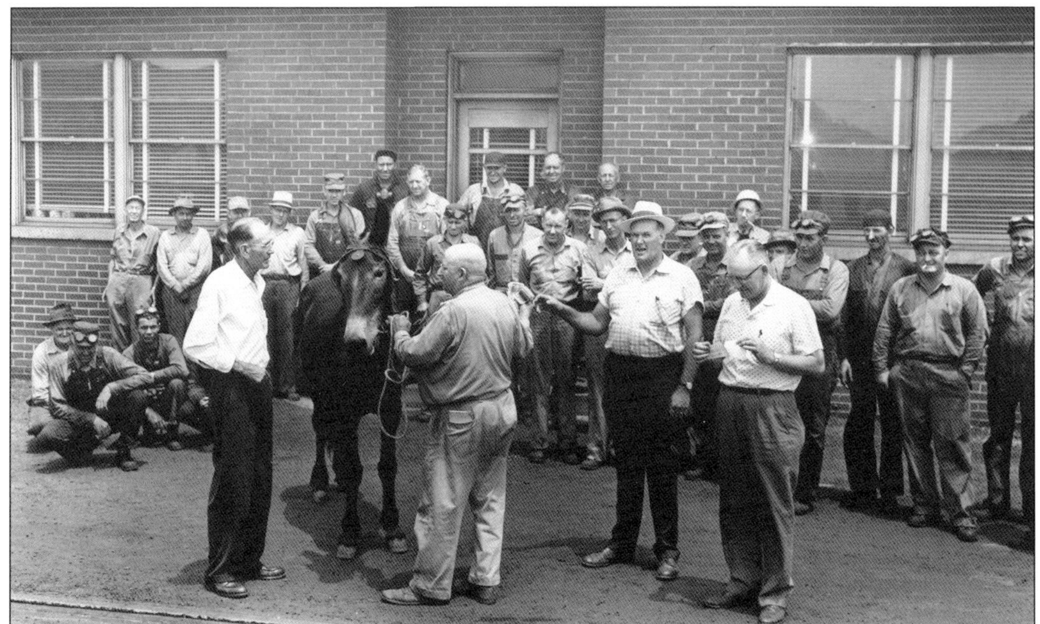

Mule Days, 1958. When smelter repairman Wilford Beavers told his coworkers that what he wanted for retirement was a mule, they took him at his word. At his retirement party, they handed him the reins to Kate, who would become his helper in the large gardening and small farming projects that were part of his plans for life after TCC.

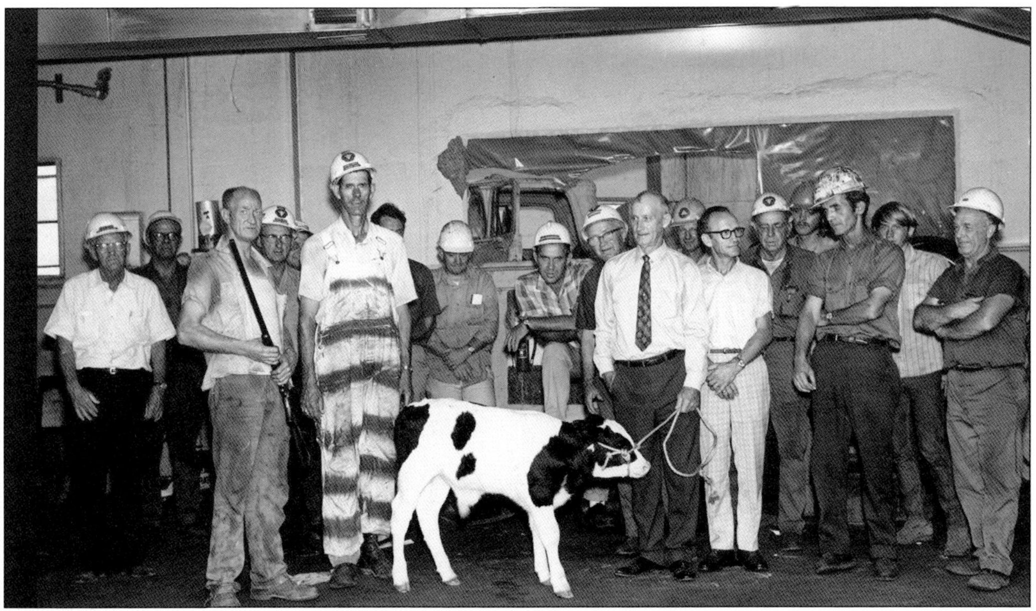

Cattle Rustlers, 1972. Like many TCC employees before him, truck driver Edgar Bryan (holding the rope) also had an aptitude for farming. At his retirement party from Cities Service in 1972, his coworkers staged a cattle rustling, complete with Harley Dillard as a rifle-carrying guard and Luther Rhodes as a prison-striped suspect, and presented him with his retirement gift: a calf.

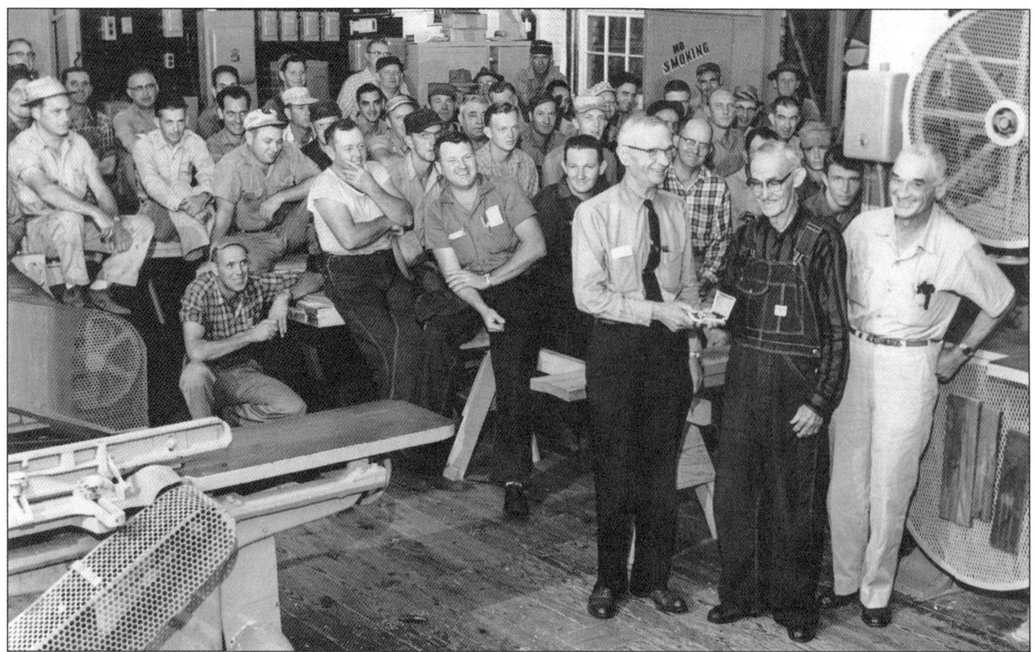

One Retirement, Two Celebrations, 1963. When mining department master mechanic Joe Lanning retired in 1963, he had worked at TCC for 54 years and had helped install every mine hoist in current operation. It was a service record that earned him not just one retirement party, but two. Above, D.O. Bigham (center) presents Lanning with a watch as a retirement gift on behalf of the Burra Burra shop crew. Below, he accepts tributes and gifts at a banquet attended by mine office staff, department supervisors, and top TCC management. Lanning's son Gene was a general foreman at the company's Isabella roasting and smelting operations, his daughter Thelma Kimsey worked as a secretary in the General Office, and his daughter-in-law Sarah Lanning worked as a chemist in the laboratory.

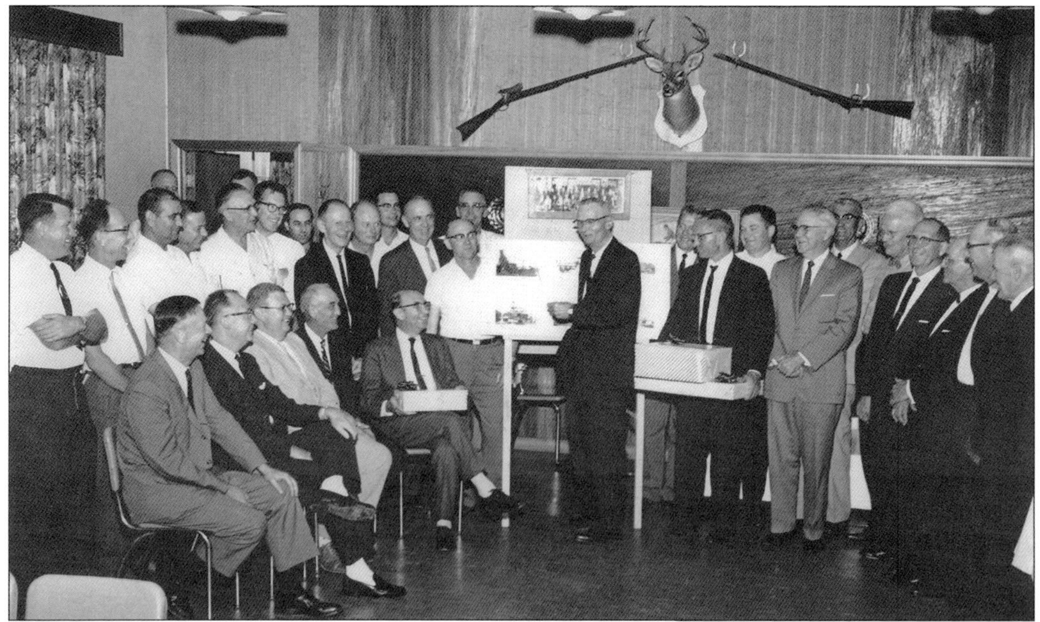

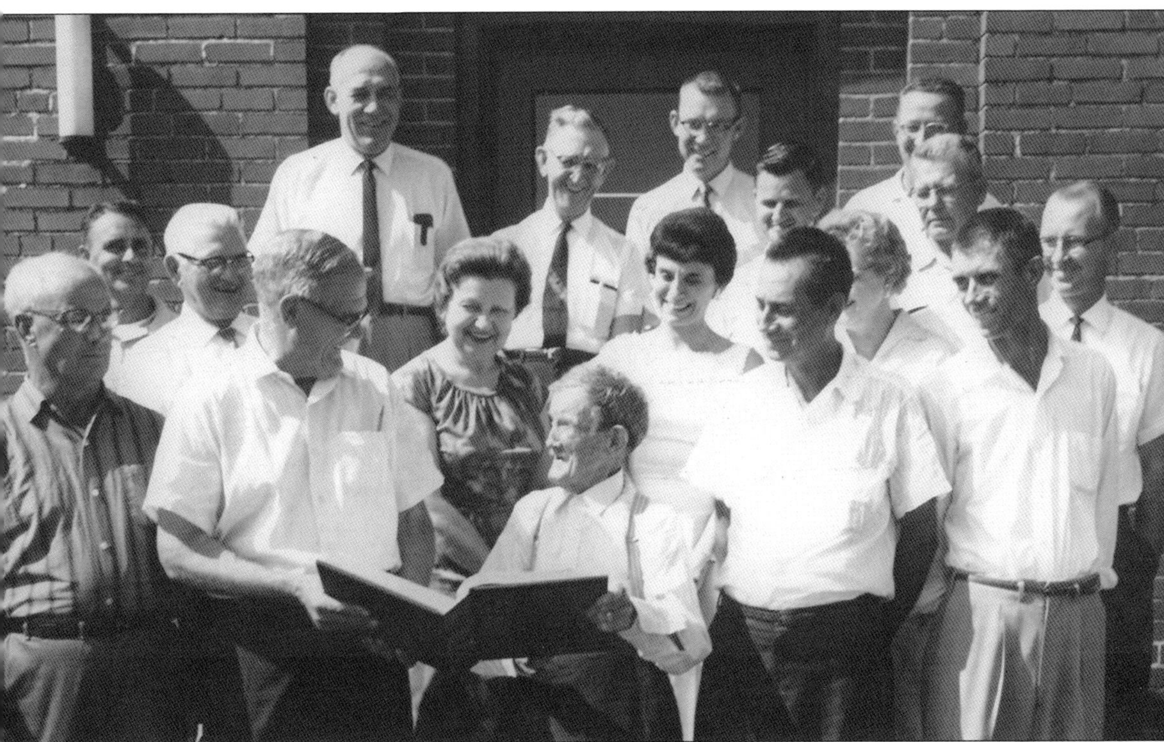

The Oldest Retiree, 1963. When TCC arrived in the Ducktown district, a young John Rider was on the spot. He remembered helping to clear the land for construction of the first TCC smelter, as well as working as a driller in the early days at Polk County Mine. He even remembered bagging a turkey in 1895 on the land where the acid plant would later stand. In 1928, he began a continuous TCC service record that ended with his retirement as a gardener at the age of 90—by far the oldest company employee at the time. In this photograph taken at his August 1963 retirement party, service office employees look on as Cliff Holden (second from left) presents him with gifts of cash and a Bible. When Rider died two years later, virtually all his male coworkers in this photograph served as pallbearers or honorary pallbearers. Also among the honorary pallbearers were TCC president T.A. Mitchell, company executives Sam Beckler and Lamar Weaver, Cowanee Club manager Vic Hensley, and company physicians H.H. "Cooney" Hyatt and W.C. Zachary.

Bibliography

Acid Test, The. Tennessee Copper Company, published from June 1943 through March 1952.

Barclay, R.E. *The Copper Basin: 1890 to 1963*. Knoxville, TN: Cole Printing & Thesis Service, 1975.

———. *Ducktown Back in Raht's Time*. Chapel Hill, NC: The University of North Carolina Press, 1946.

Science Applications International Corporation. *History of Tennessee Copper Company and Successor Firms at the Copperhill Plant and the Ducktown Mining District, Copper Basin, Tennessee*. Prepared for the US Army Corps of Engineers, Nashville District, and the US Environmental Protection Agency, Region 4, 2008.

TC Topics. Tennessee Copper Company, published from April 1952 through December 1969.

The Ducktown Basin Museum

The Ducktown Basin Museum is located on the site of the historic Burra Burra Mine, which was the headquarters for Tennessee Copper Company and Cities Service mining operations from 1899 through 1975. In 1981, Cities Service donated the site to the museum, then in its infancy and housed in a small storefront location on Main Street in Ducktown. The museum formally took over administration of the property in early 1982.

The site, now owned by the State of Tennessee and operated as a state historic site under an agreement between the museum and the Tennessee Historical Commission, has been listed in the National Register of Historic Places since 1983. The 16 remaining structures include virtually all of the original mine buildings and outbuildings except for the Burra Burra headframe, which was demolished after the closing of the mine in 1958. The mine office houses the museum's collection and is the only building permanently open to visitors. A self-guided walking tour of the grounds is available during daylight hours.

With the closing of the mines in 1987, the museum was able to acquire additional artifacts from the local mining operations. Today's visitors can wander among the giant drills and wrenches, surveyors' instruments, portable telephones, signal code signs, and shift whistles that were part of the miners' everyday lives. Indoor exhibits, which include both artifacts and photographs, interpret the mining, industrial, and cultural history of the Ducktown district. Larger artifacts are displayed on the museum grounds. The gift shop features a variety of merchandise that includes handcrafted items, souvenirs, and books and DVDs of local interest.

The museum is open year-round, with hours varying by season. More information is available at www.ducktownbasinmuseum.com.